T0144607

Ripple-Down Rules

Ripple-Down Rules

The Alternative to Machine Learning

Paul Compton
The University of NSW, Sydney, Australia

Byeong Ho Kang
The University of Tasmania, Hobart, Australia

CRC Press
Taylor & Francis Group
Boca Raton London New York

CRC Press is an imprint of the
Taylor & Francis Group, an **informa** business
A CHAPMAN & HALL BOOK

First edition published 2021
by CRC Press
6000 Broken Sound Parkway NW, Suite 300, Boca Raton, FL 33487-2742

and by CRC Press
2 Park Square, Milton Park, Abingdon, Oxon, OX14 4RN

© 2021 Paul Compton, Byeong Ho Kang

CRC Press is an imprint of Taylor & Francis Group, LLC

The right of Paul Compton, Byeong Ho Kang to be identified as authors of this work has been asserted by them in accordance with sections 77 and 78 of the Copyright, Designs and Patents Act 1988.

Reasonable efforts have been made to publish reliable data and information, but the author and publisher cannot assume responsibility for the validity of all materials or the consequences of their use. The authors and publishers have attempted to trace the copyright holders of all material reproduced in this publication and apologize to copyright holders if permission to publish in this form has not been obtained. If any copyright material has not been acknowledged please write and let us know so we may rectify in any future reprint.

Except as permitted under U.S. Copyright Law, no part of this book may be reprinted, reproduced, transmitted, or utilized in any form by any electronic, mechanical, or other means, now known or hereafter invented, including photocopying, microfilming, and recording, or in any information storage or retrieval system, without written permission from the publishers.

For permission to photocopy or use material electronically from this work, access www.copyright.com or contact the Copyright Clearance Center, Inc. (CCC), 222 Rosewood Drive, Danvers, MA 01923, 978-750-8400. For works that are not available on CCC please contact mpkbookspermissions@tandf.co.uk

Trademark notice: Product or corporate names may be trademarks or registered trademarks and are used only for identification and explanation without intent to infringe.

Library of Congress Cataloging-in-Publication Data
Names: Compton, Paul, 1944- author. | Kang, Byeong-Ho, author.
Title: Ripple-down rules : the alternative to machine learning / Paul Compton, The University of NSW, Sydney, Australia, Byeong Ho Kang, The University of Tasmania, Hobart, Australia. Description: First edition. | Boca Raton : CRC Press, 2021. | Includes bibliographical references and index.
Identifiers: LCCN 2020047019 | ISBN 9780367644321 (paperback) | ISBN 9780367647667 (hardback) | ISBN 9781003126157 (ebook)
Subjects: LCSH: Expert systems (Computer science) | Ripple down rules (Machine learning) Classification: LCC QA76.76.E95 C653 2021 | DDC 006.3/3–dc23
LC record available at https://lccn.loc.gov/2020047019

ISBN: 978-0-367-64766-7 (hbk)
ISBN: 978-0-367-64432-1 (pbk)
ISBN: 978-1-003-12615-7 (ebk)

Typeset in Minion Pro
by SPi Global, India

Contents

Preface

Artificial Intelligence (AI) once again seems to hold out the promise of doing extraordinary things, particularly through the magic of machine learning. But to do extraordinary things an AI system needs to have a lot of knowledge. For example:

- Self-driving cars have to understand what their vision system, and their other sensors, are telling them about the world and they have to know what to do in all the circumstances they will encounter. A huge amount of knowledge about the world and how to interact with it is fundamental to being able to drive well.
- If a medical AI system which provides advice to GPs about laboratory results is going to provide detailed patient-specific advice at the level of a consultant chemical pathologist, it has to know not only about the whole range of diagnoses in relation to laboratory results but also how these are affected by drugs, patient history and so on – and how the results impact on-going patient management.

The current focus is on seeking to acquire such knowledge through machine learning, but as will be discussed, datasets of sufficient quality to learn such detailed knowledge are problematic. Alternatively, one can also seek to obtain knowledge from an expert, through a process of knowledge elicitation and engineering. The difficulties in doing this will also be discussed, but Ed Feigenbaum's comment over 30 years ago still applies.

> "The problem of knowledge acquisition is the critical bottleneck problem in artificial intelligence." (Feigenbaum 1984)

One approach to the problem of acquiring knowledge from people is known as Ripple-Down Rules (RDR) and is the focus of this book.

RDR has had significant practical success. For example, IBM investigated using RDR to data cleanse Indian street address data, apparently a difficult problem (Dani et al. 2010). Their RDR method outperformed the machine learning methods and the commercial system they also investigated, and the two main researchers received an IBM award granted if a piece of research leads to more than $10M of new business. Pacific Knowledge Systems (PKS)[1] provides RDR technology for chemical pathology laboratories to provide patient-specific interpretations of lab results. Worldwide, there are over 800 PKS RDR knowledge bases deployed, ranging from 100s to over 10,000 rules. These are developed by pathologists themselves after a couple of day's training. They build a system while it is in use, adding a rule whenever they notice that one of the patient reports they monitor as part of their duties has been given an inappropriate interpretation by the evolving knowledge base. It takes on average a couple of minutes for them to add and validate a new rule. To our knowledge, no other knowledge-based system technology has achieved this level of empowering users. Occasionally domain experts do build knowledge bases themselves using other technologies – but essentially by becoming knowledge engineers, whereas with RDR, rule building is a minor extension to their normal duties.

Given results like this, one would expect substantial industry uptake of RDR technology and there are at least eight companies using RDR, but all but one of these had some personal connection with another RDR project elsewhere. Why has this personal connection and direct experience been needed?

We suspect that the reasons are firstly that RDR can seem quite counter-intuitive. As discussed in Chapter 2, it is based on a different philosophical understanding from most other AI on what knowledge is. In particular, a key principle in every AI textbook and university AI course in the last 40–50 years has been that knowledge and reasoning have to be separate. This is particularly the case for knowledge-based systems with their separation of the knowledge base and the inference engine. In contrast, RDR explicitly rejects this separation and the knowledge base itself specifies the order of evaluation of rules. Perhaps even more counter-intuitive to an engineer is that there is no requirement to structure the knowledge; it is just added over time as cases require rules. In this book we make the further seemingly counter-intuitive suggestion that building an RDR system is probably going to be cheaper and perform better than a machine-learning-based system if the data labels for the training data are based on human assessment – hence the title of this book.

Professor Ashwin Srinivasan from BITS Pilani worked on PEIRS, the first large deployed RDR system, and introduced RDR to IBM when he worked for IBM research. He has repeatedly suggested to us that what is needed is not more academic papers, but an RDR manual so that people could experience for themselves how effective RDR is. This book is the response to Ashwin's suggestion.

The book is structured as follows:

Chapters 1 and 2 discuss why an RDR approach is needed, covering the problems with both machine learning and knowledge acquisition from experts. Readers who want to get straight into the technology might want to skip these chapters, but since RDR is based on different philosophical assumptions about knowledge, these chapters may be important for appreciating: why RDR?

Chapters 3–7 are essentially the manual for various types of RDR. Some readers will want to move quickly to using the open-source Excel_RDR tools to understand RDR, but on the other hand the worked examples in these chapters should provide sufficient detail to understand exactly how RDR works.

Chapter 8 provides some implementation advice, particularly relating to validation.

Chapter 9 revisits machine learning and how RDR can be used with machine learning or as an alternative.

Appendices 1 and 2 outline the various applications where RDR has been used in industry or demonstrated in research.

Currently the Excel_RDR software used in Chapters 5 to 7 and the associated manual can be downloaded only from http://www.cse.unsw.edu.au/~compton/RDR_software.html. It will also become available on general download sites.

Finally, we might note that RDR has another perhaps more political difference from other AI technology. We live in an age where people are very concerned that AI is going to intervene increasingly in their lives. RDR is not so much an AI, an artificial intelligence, but an IA or IB, an intelligence amplifier or broadcaster. RDR systems are built and completely controlled by their user to do exactly what the user wants them to do and can be endlessly adapted as the user requires. Essentially, what

RDR does is empower users to do their job better, to take over the more boring repeated exercise of their expertise. Knowledge-based systems have always held out this hope – but RDR does this in a way that empowers the user, rather than replacing them.

NOTE

1 https://pks.com.au Paul Compton was a co-founder of PKS and until 2019 a minority shareholder.

Acknowledgements

Ripple-Down Rules (RDR) would never have emerged except that the then director of the Garvan Institute, Les Lazarus AO, encouraged, supported and sponsored the development of one of the first medical expert systems to go into routine use, GARVAN-ES1, which exposed the maintenance problems of expert systems. Later he sponsored PEIRS, the first RDR system to go into routine clinical use and co-founded and chaired Pacific Knowledge Systems, the first RDR company.

RDR might have been only an application exercise without the encouragement of Brian Gaines and Mildred Shaw, through the European and Banff Knowledge Acquisition Workshops, that RDR were an important contribution to knowledge acquisition research. Brian also pioneered machine learning using an RDR representation.

Ashwin Srinivasan implemented the shell for PEIRS, the first medical RDR system to go into clinical use and this book would not have been written except that he finally persuaded us, based on his later experience, that a manual rather than more academic papers would better promote RDR in industry. Glenn Edwards was the chemical pathologist who wrote the rules and developed the knowledge base for PEIRS.

The initial success of Pacific Knowledge Systems (PKS) in applying RDR has been critical in demonstrating the utility of the approach. The founders who took the risk in establishing PKS were Les Lazarus, Glenn Edwards, Jeffrey Braithwaite, Richard Brown, Lindsay Peters and Paul Compton. Lindsay Peters, PKS chief technology officer, is the only one still connected with PKS. We are particularly grateful to Lindsay for many discussions over the years and his many insights about RDR based on PKS's experience.

Finally, the material in this book has come from the work of the many people who have developed and explored the possibilities for RDR. There are too many names to list here, but they are all (we hope) listed, along with their contributions to RDR development, in the appendices or bibliography. In particular Hiroshi Motoda, then at Hitachi Research Labs, created a postdoc for Byeong Kang to pursue further research on RDR after his PhD and later supported RDR research funding through his role at the US Airforce, Asian Office of Aerospace Research and Development. We also thank him for his helpful comments on this manuscript.

About the Authors

Paul Compton initially studied philosophy before majoring in physics. He spent 20 years as a biophysicist at the Garvan Institute of Medical Research, and then 20 years in Computer Science and Engineering at the University of New South Wales, where he was head of school for 12 years and is now an emeritus professor.

Byeong Ho Kang majored in mathematics in Korea, followed by a PhD on RDR at the University of New South Wales and the algorithm he developed is the basis of most industry RDR applications. He is a professor, with a research focus on applications, and head of the ICT discipline at the University of Tasmania.

CHAPTER 1

Problems with Machine Learning and Knowledge Acquisition

1.1 INTRODUCTION

Ripple-Down Rules (RDR) are intended for problems where there is insufficient data for machine learning and suitable data is too costly to obtain. On the other hand, RDR avoids the major problems with building systems by acquiring knowledge from domain experts. There are various types of RDR, and this book presents three of them. Although RDR is a knowledge acquisition method acquiring knowledge from people, it is perhaps more like machine learning than conventional knowledge engineering.

In the 1970s and early 80s there were huge expectations about what could be achieved with expert or knowledge-based systems based on acquiring knowledge from domain experts. Despite the considerable achievements with expert systems, they turned out to be much more difficult to build than expected, resulting in disillusionment and a major downturn in the new AI industry. We are now in a new phase of huge expectations about machine learning, particularly deep learning. A 2018 Deloitte survey found that 63% of the companies surveyed were using machine learning in their businesses with 50% using deep learning (Loucks, Davenport, and Schatsky 2018). The same survey in 2019 shows a small increase in the use of machine learning over 2018, but also that 97% of respondents plan to use machine learning and 95% deep learning, respectively, in the next year (Ammanath, Hupfer, and Jarvis 2019). It appears that machine learning is a magic new technology whereas in fact the history of machine learning goes back to the early 1950s when the first neural network machine was developed based on ideas developed in the 1940s. The first convolutional neural networks, a major form of deep learning, were developed in the late 1970s. Although this is the first book on Ripple-Down Rules, RDR also has some history. An RDR approach to address the maintenance challenges with GARVAN-ES1, a medical expert system, was first proposed in 1988 (Compton and Jansen 1988) only three years after GARVAN-ES1 was first reported (Horn et al. 1985) and two years after GARVAN-ES1 was reported as one of the first four medical expert systems to go into clinical use (Buchanan 1986). The reason for an RDR book now is to present a fall-back technology as industry becomes increasingly aware of the challenges in providing data good enough for machine learning to produce the systems they want. We will first look briefly at the limitations and problems with machine learning and knowledge acquisition.

1.2 MACHINE LEARNING

Despite the extraordinary results that machine learning has produced, a key issue is whether there is sufficient reliably labelled data to learn the concepts required. Despite this being the era of big data, providing adequate appropriate data is not straightforward. If we take medicine as an example: A 2019 investigation into machine learning methods for medical diagnosis identified 17 benchmark datasets (Jha et al. 2019). Each of these has at most a few hundred cases and a few classes, with one dataset having 24 classes. This sort of data does not represent the precision of human clinical decision making. We will later discuss knowledge bases in Chemical Pathology which are used to provide expert pathologist advice to clinicians on interpreting patient results. Some of these knowledge bases provide hundreds, and some even thousands of different conclusions. Perhaps Jha et al.'s investigation into machine learning in medicine did not uncover all the datasets available, but machine learning would fall far short of being able to provide hundreds of different classifications from the datasets they did identify.

Hospitals receive funding largely based on the discharge codes assigned to patients. A major review of previous studies of discharge coding accuracy found the median accuracy to be 83.2% (Burns et al. 2011). More recent studies in more specialised, and probably, more difficult domains show even less accuracy (Ewings, Konofaos, and Wallace 2017, Korb et al. 2016). Chavis provides an informal discussion of the problems with accurate coding (Chavis 2010). No doubt discharge coding has difficulties, but given that hospital funding relies on it, and hospitals are motivated to get it right, it appears unlikely that large data bases with sufficient accuracy to be used by machine learning for more challenging problems are going to be available any time soon.

At the other end of the scale we have had all sorts of extraordinary claims about how IBM's Watson was going to transform medicine by being able to learn from all published medical findings. It was the ultimate claim that given the massive amount of information in medical journal articles and implicit in other data, machine learning should be able to extract the knowledge implicit in these data resources. Despite Watson's success playing Jeopardy, this has not really translated to medicine (Strickland 2019). For example, in a study of colon cancer treatment advice in Korea, Watson's recommended treatment only agreed with the multidisciplinary team's primary treatment recommendations 41.5% of the time, but it did agree on treatments that could be considered, 87.7% of the time (Choi et al. 2019). It was suggested that the discordance in the recommended treatments was because of different medical circumstances between the Korean Gachon Gil Medical Centre and the Sloan Kettering Cancer Centre. This further highlights a central challenge for machine learning: that what is learned is totally dependent on the quality and relevance of the data available. There is also the question of how much human effort goes into developing a machine learning system. In IBM's collaboration with WellPoint Inc. 14,700 hours of nurse-clinician training were used as well as massive amounts of data (Doyle-Lindrud 2015). The details of what this training involved are not available, but 6–7 man-years of effort is a very large effort on top of the machine learning involved. This collaboration led to the lung cancer program at the Sloan Kettering Cancer Centre using Watson; however, recent reports of this application indicate that for the system used at the Sloan Kettering, Watson was in fact trained on only hundreds of synthetic cases developed by one or two doctors and its recommendations were biased because of this training (Bennett 2018). Data on the time taken to develop these synthetic cases does not seem to be available. If one scans the Watson medical literature, the majority of the publications are about the potential of the approach, rather than results. There is no doubt that the Watson's approach has huge potential and will eventually achieve great things, but it is also clear that the results so far have depended on a lot more than just applying learning to data – and have a long way to go to match expert human judgement.

This central issue of data quality was identified in IBM's 2012 Global Technology Outlook Report (IBM Research 2012) naming "Managing Uncertain Data at Scale" as a key challenge for analytics and learning. A particular issue is the accuracy of the label or classification applied to the data, as shown in the discharge coding example above. If a label attached to a case is produced automatically, it is likely to be produced consistently and the dataset is likely to be highly useful. For example, if data on the actual outcome from an industrial process is available as well as data from sensors used in the process, then the data should be excellent for learning. In fact, one of the earliest successful applications of machine learning was for a Westinghouse fuel sintering process where a decision tree algorithm discovered the parameter settings to produce better pellets, boosting Westinghouse's 1984 income by over $10M per year (Langley and Simon 1995). Apparently, the system outperformed engineers in predicting problems. The ideal application for machine learning is not only when there is a large number of cases, but where the label or classification attached to the case is independent of human judgement; e.g. the borrower did actually default on their loan repayment, regardless of any human assessment.

Human biases in making judgements are well known (Kahneman, Slovic, and Tversky 1982), but we are also inconsistent in applying labels to data. In a project where expert doctors had to assess the likelihood of kickback payments from pathology companies to GPs, the experts tended to be a bit inconsistent in their estimates of the likelihood of kickback. However, if they were constrained to identify differentiating features to justify a different judgement about a case, they became more consistent (Wang et al. 1996). It so happened that RDR were used to ensure they selected differentiating features, but the point for the discussion here is simply that it is difficult for people to be consistent in subtle judgements. As will be discussed, human judgement is always made in a context and may vary with the context.

One approach to human labelling of large datasets is crowdsourcing, but careful attention has to be paid to quality. Workers can have different levels of expertise or may be careless or even malicious, so there have to be ways of aggregating answers to minimise errors (Li et al. 2016). But clearly crowdsourcing has great value in areas such as labelling and annotating images, and when deep learning techniques are applied to such datasets extremely good results are achieved, which could not be achieved any other way.

An underlying question is: how much data and what sort of data is needed for a machine learning algorithm to be able to do as well as human judgement. An IBM project on data cleansing for Indian street address data provides an example of this issue (Dani et al. 2010). Apparently, it is very difficult to get a clean dataset for Indian street addresses. The methods used in this study were RDR, a decision tree learner, a conditional random field learner and a commercial system. The version of RDR used, with different knowledge bases for each of the address fields and the use of dictionaries, was more sophisticated (or rather specialised) than the RDR systems described in this book. The systems developed were trained on Mumbai data and then tested on other Mumbai data and data from all of India.

TABLE 1.1 Precision of various methods to learn how to clean Indian street address data. This table has been constructed from data presented in (Dani et al. 2010)

Method	no of training examples	Mumbai test data	All India test data
RDR	60	75%	68%
Decision Tree	600	75%	45%
CRF	600	80%	40%
Commercial		60%	48%

As seen in Table 1.1 all the methods, except for the commercial system, performed comparably when tested on data similar to the data on which they were trained, with a precision of 75–80%. However, when tested on data from all of India, although all methods degrade, the RDR method degrades much less than the statistical methods. The issue is not so much the use of RDR, but that a method using human knowledge, based on years of lived experience, is likely to do better than purely statistical methods – unless they have a lot of data to learn from.

If more training data had been available than the 600 cases used, no doubt the machine learning methods would have done better, but the question arises: where does this data comes from? To create training data, the correct address has to be matched with the ill-formed addresses and presumably people do this matching. If this matching could have been automated there would have been no need for this research which led to commercial application. If people have to be used to do the labelling, then why not get the same people to write rules as they go. This is a central motivation for an RDR approach and if data from Pacific Knowledge

Systems customers (see Appendix 1) is typical, it will take them only a couple of minutes or less to write a rule. This leads to the perhaps counter-intuitive suggestion that if data for machine learning has to be labelled by people, then there may well be less human effort required in building an RDR knowledge base than in labelling sufficient cases for a machine learner to produce a knowledge base of the same quality. This recalls Achim Hoffmann's paper on the general limitations of machine learning (Hoffmann 1990). Hoffman argued that people have to do the same work whether they write a complex program or provide a learner with sufficient data; that is, you can't expect magic, you either write the program or provide sufficient data where the information needed is embedded – there is no free lunch. Of course, there are short cuts, if e.g. data can be labelled automatically as in the Westinghouse example. On the other hand, should we expect there will always be some cases which are so rare that it is almost impossible to get sufficient examples? But a human expert will know exactly what conclusion or label should apply to this data in the context, and why.

All of this discussion leads to the conclusion that despite the power of machine learning and its wide application, providing sufficient high-quality data for a learner can be a difficult problem and it may perhaps be simpler to incorporate human knowledge into a program – an expert or knowledge-based system. But it was precisely because of the difficulties in incorporating human knowledge into knowledge-based systems that machine learning has come so much to the fore!

What we have been discussing is supervised learning: machine learning where the data includes the label that is to be learned. This is where learning from domain experts is relevant as they have the expertise to assign labels to case data in their domain, e.g. in medical diagnosis. There are also other forms of machine learning, with the furthest from supervised learning being unsupervised learning where the data does not have any labels. Obviously, you can't learn labels if the data doesn't have labels, but you can learn how to cluster or simplify data in some way, which can be very useful. If we assume there are patterns in the data rather than it being completely random, an autoencoder can be used to learn a reduced encoding of the data according to some appropriate criteria. For example, back-propagation deep learning can learn a mapping between 2D images and 3D human poses (Hong et al. 2015). What is being learned with these types of learners are overall patterns in the data rather than a label or classification for a particular case or data instance.

But even if we can provide all the high-quality data needed for supervised machine learning, there remains one last issue that is of increasing importance for both supervised and unsupervised learning. The power of deep learning is that the neural network layers are able to discover implicit features, and this is where its learning power comes from. But these "features" are not part of human language, so how can a deep learner explain and justify its decisions? One obvious area of concern is in policing. If you are using deep learning to identify possible criminal intent, do you end up just targeting the marginalised? If you use deep learning for credit scoring do you again end up automatically giving a lower score to the more marginalised? These are critical problems and since 2018 the European Union General Data Protection Regulation has required that AI or other systems should be able to explain their decisions, and these should be able to be challenged. Perhaps purely algorithmic approaches will be able to produce the required explanations, but it seems more likely that some sort of combining with human knowledge will be required.

1.3 KNOWLEDGE ACQUISITION

Machine learning has become so prominent largely because of the difficulty of incorporating human knowledge into a knowledge base. The phrase "the knowledge engineering bottleneck" has been used since the 1980s and was probably introduced by Ed Feigenbaum as it is also referred to as "the Feigenbaum bottleneck". It was assumed that since rules were modular and independent of the inference engine, acquiring knowledge should have been a simple matter of domain experts providing rules. This has never been the case; Bobrow et al. (Bobrow, Mittal, and Stefik 1986) in their survey of three well-known early systems, R1, Pride and the Dipmeter advisor concluded:

> *Expert Knowledge Has to Be Acquired Incrementally and Tested.* Expert knowledge is not acquired all at once: The process of building an expert system spans several months and sometimes several years. In the course of this development, it is typical to expand and reformulate the knowledge base many times. In the beginning, this is because choosing the terminology and ways of factoring the knowledge base is subject to so much experimentation. In the middle phases, cases at the limits of the systems capabilities often expose the need to reconsider basic categories and organization. Approaches viable for a small knowledge base

> and simple test cases may prove impractical as larger problems are attempted. . . . . Toy programs for a small demonstration can be built quickly-often in just a few months using current technology. However, for large-scale systems with knowledge spanning a wide domain, the time needed to develop a system that can be put in the field can be measured in years, not months, and in tens of worker-years, not worker-months.

Matthew Fox in "AI and expert system myths, legends and facts" (Fox 1990) identified the difficulty as follows:

> *LEGEND: AI systems are easy to maintain.* Using rules as a programming language provides programmers with a high degree of program decomposability; that is, rules are separate knowledge chunks that uniquely define the context of their applicability. To the extent that we use them in this manner, we can add or remove rules independently of other rules in the system, thereby simplifying maintenance. Building rule-based systems differs from this ideal. Various problem-solving methods (including iteration) require that rules implementing these methods have knowledge of other rules, which breaks the independence assumption and makes the rule base harder to maintain. The much-heralded XCON system has reached its maintainability limit (about 10,000 rules). The complexity of rule interactions at this level exceeds maintainer abilities.

Zacharias' survey of modern rule-system developers (Zacharias 2008) came to similar conclusons 18 years later. 64 of the 76 respondents answered most questions with reference to the largest knowledge base they had worked on in the last five years. The respondents had over 6.6 years' experience developing knowledge-based systems and used a range of rule technologies. 60% of the respondents indicated that their knowledge bases *frequently* failed to give the correct result and 34% indicated that *sometimes* the incorrect results were given. The biggest need was identified as debugging tools to correct such errors – confirming the observation of Bobrow et al. 22 years, and Fox 18 years, earlier – that it is tedious and messy to build a sophisticated knowledge base and one needs to painstakingly test and fix rule interactions.Elsewhere Zacharias wrote:

> *The One Rule Fallacy:* Because one rule is relatively simple and because the interaction between rules is handled automatically by the inference engine, it is often assumed that a rule base as a whole is automatically simple to create. . . . . . However, it is an illusion to assume that rules created in isolation will work together automatically in all situations. Rules have to be tested in their interaction with other rules on as many diverse problems as possible to have a chance for them to work in novel situations.
>
> (Zacharias 2009)

In the 1990s the dominant approach to improving knowledge engineering was to move away from the notion of obtaining knowledge and rather consider the domain expert and knowledge engineer as collaborating in building a problem-solving model for the domain. The best-known example of this approach is probably CommonKADS (Schreiber et al. 1994, Schreiber et al. 1999, Speel et al. 2001), essentially a comprehensive software engineering framework for building knowledge-based systems. Despite the obvious value in such systematic approaches, the same problem of debugging and maintenance remains. As the CommonKADS authors themselves note:

> Although methodologies such as CommonKADS support the knowledge acquisition process in a number of ways (e.g. by providing modelling constructs and template models) experience shows that conceptual modelling remains a difficult and time-consuming activity.
>
> (Speel et al. 2001)

Despite these clear statements that knowledge-base debugging and maintenance is and has always been a major problem, there seem to be few case studies documenting maintenance problems. Wagner in a longitudinal survey of 311 published expert system case studies did not find anything on maintenance (Wagner 2017). This failure to report on maintenance problems is perhaps because researchers tend to write up successful expert system developments early on. The only two long-term maintenance reports we are aware of are the reports on XCON, one of the landmark developments in knowledge-based systems and the much smaller

GARVAN-ES1, which nevertheless was reviewed as one of the first four medical expert systems to go into routine clinical use (Buchanan 1986).

XCON was used to configure DEC VAX computers against customer requirements and was the outstanding expert system in industry use in the early years of expert systems. The initial system was developed in Carnegie-Mellon University (CMU), but deploying XCON at DEC involved over a year of training for DEC engineers in how to maintain XCON, then with about 1,000 rules. These maintenance demands meant engineers were unable to also maintain other expert systems introduced from CMU to DEC (Polit 1984). XCON eventually had 6,500 rules with 50% changed every year, as new products and versions were introduced – a major maintenance challenge (Soloway, Bachant, and Jensen 1987). Apparently 40 programmer/knowledge engineers were required to maintain XCON (Sviokla 1990) and as noted by Fox the limit of maintainability for XCON was probably about 10,000 rules (Fox 1990). XCON was built using essentially the same technology as modern expert systems are based on, the OPS RETE architecture (Forgy and McDermott 1977).

GARVAN-ES1 was a small medical expert system providing interpretative comments for thyroid laboratory reports (Horn et al. 1985). The purpose of a comment appended to a report of thyroid test results was to advise the referring GP on the clinical interpretation of the results. As of 1989, GARVAN-ES1 had 276 rules, but it also allowed disjunctions and there were 262 disjunctions suggesting about 500–600 rules if disjunctions were disallowed, and this approximates later experimental rebuilds. GARVAN-ES1 was put into clinical use after a standard evaluation on unseen clinical data showing experts agreed with its conclusions 96% of the time. However, since the performance of expert systems in actual clinical practice was such an unknown (GARVAN-ES1 being one of the first four medical expert systems in clinical use(Buchanan 1986)), all reports produced by the system were checked by one of the endocrine registrars and the rules updated whenever a registrar was not happy[1] with the comment made by GARVAN-ES1. The rules were constantly edited and updated to provide the required comments, except where the change requested seemed to the knowledge engineer to be too minor and this was confirmed with the Garvan Institute's senior endocrinologist.

The GARVAN-ES1 maintenance experience is illustrated in Figure 1.1 and shows the changing size of the knowledge base over four years maintenance. The knowledge base size is shown in kilobytes rather than numbers of rules as rules contained disjunctions. Over four years the knowledge

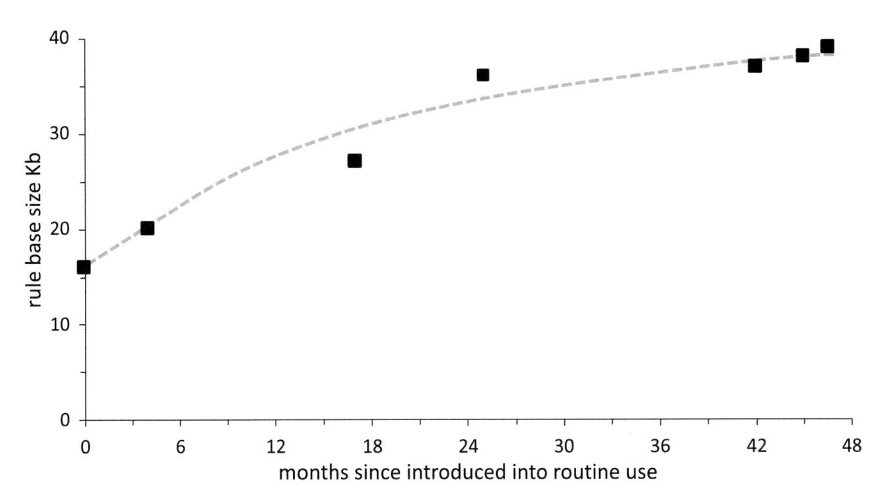

FIGURE 1.1 The increasing size of the GARVAN-ESI knowledge base (redrawn from Compton and Jansen 1990).

base doubled in size while the accuracy went from 96% to 99.7%. Perhaps a lot of these changes were not really necessary and the registrars involved wanted a level of clinical detail in the comments that was not really necessary; nevertheless whether the changes were trivial or important they were made to enable the system to do what the users expected it to do.

Another view of these changes is shown in Figure 1.2. Every time a rule was changed the case that prompted the change was stored as a "cornerstone" case – that is, a case for which rules have been added or changed.

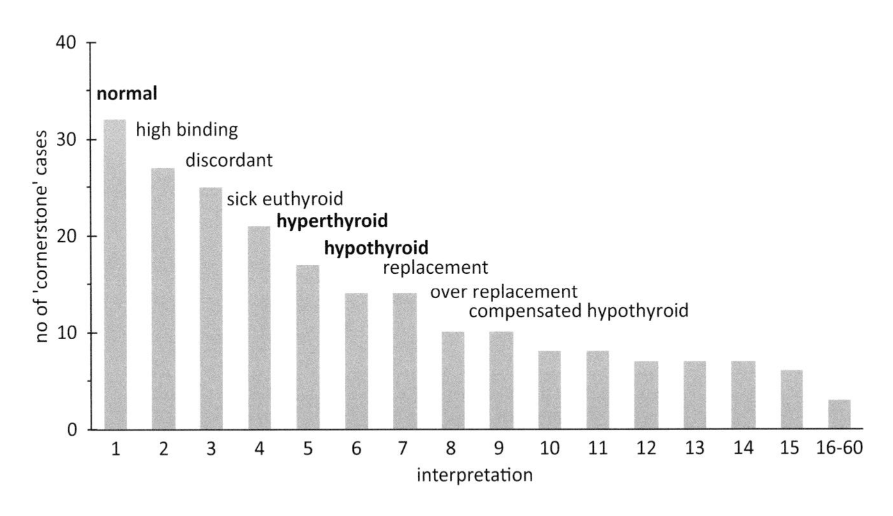

FIGURE 1.2 The number of cornerstone cases for each conclusion (redrawn from Compton et al. 1988).

Figure 1.2 shows the number of cornerstone cases that had been seen for the 60 different interpretations given (59 comments plus no comment as the default for normal results). The number of cases for each interpretation is shown only for the 15 interpretations with the greatest number of cases and the average number of cornerstone cases is shown for interpretations 16–60. The interpretation "toxic" is classic primary hyperthyroidism while "hypothyroid" is classic primary hypothyroidism. The key diagnostic features of primary hyperthyroidism are trivial and well known: elevated thyroid hormones and suppressed thyroid stimulating hormone and conversely the key diagnostic features for primary hypothyroidism are elevated thyroid stimulating hormone and low thyroid hormones. Despite the apparent clarity of the reasons for these diagnoses, Figure 1.2 shows that during the evolution of the GARVAN-ES1 knowledge base, errors were made on 21 toxic (primary hyperthyroidism) cases and 17 primary hypothyroidism cases which required rules to be corrected. As well, 32 normal cases were misdiagnosed as the knowledge base evolved. Superficially, these are very surprising results and provide a classic example of the so-called knowledge engineering bottleneck – experts don't readily tell the knowledge engineer everything that the knowledge base needs to know. This isn't a lack of expertise on the expert's part, the expert is perfectly capable of correctly interpreting any case they are presented with, but this is quite different from providing rules which will cover other unseen cases.

We believe that the problems in knowledge engineering, ultimately, are not related to any failure of experts to report on their mental processes or the knowledge they use, but are because our expectations and assumptions about knowledge are mistaken – we have the wrong philosophical assumptions as will be discussed in Chapter 2.

In the discussion above we have assumed that the domain expert can articulate their knowledge or explain their decision about a case in terms that can be easily communicated to a computer. For example, the expert might refer to: *age > 75*, *temperature is increasing*, *email subject heading contains the word 'crisis'* etc. All of these can be fairly readily coded for a computer. The knowledge acquisition problem we have discussed is the difficulty in getting an expert to provide sufficient and appropriate knowledge of this type to develop a truly expert system.

There is a quite different problem that in some domains experts simply do not provide their knowledge in a way that be coded. For example, a radiologist may look at a lung X-ray and justify their decision by referring to features in the image such as *ground-glass opacity* or *honeycombing*. These features are a very long way from the pixel data that comprises the computer's version of the X-ray image. The knowledge acquisition problem we have discussed still remains of whether the expert has articulated sufficient and appropriate knowledge in terms of features like *ground-glass opacity* and *honeycombing*, but there is the separate and different issue of whether the expert's language for describing features can be coded.

The discussion in this book is restricted to application problems where the feature referred to by a domain expert can be readily coded. We suggest that this is in fact the large majority of applications, but there are other important applications such as image processing, where methods like deep learning are more appropriate, because the expert's features are too high level compared to the actual data stored. But such methods still face the problem of sufficient accurately labeled data.

NOTE

1 A registrar disagreeing with a comment did not necessarily mean it was incorrect or clinically dangerous; it could simply mean that the registrar wanted to draw attention to some particular features in the data.

CHAPTER 2

Philosophical Issues in Knowledge Acquisition[1]

One of the central ideas underpinning research in Artificial Intelligence (AI), particularly rule-based systems, is that both machines and people are symbol-manipulating systems and that this is the essence of intelligence. In his 1975 Turing Award Lecture, Newell outlined his and Simon's physical symbol hypothesis:

> A physical symbol system has the necessary and sufficient means for general intelligent action.
>
> (Newell and Simon 1976)

There can be no argument that we are symbol-manipulating systems, but does this mean we have a store of symbols and their logical relations in our minds/brains that we use for intelligent decision making?

The language often suggests an implicit belief that we do have a store of knowledge in our mind/brains and the knowledge acquisition bottleneck is because of the difficulty of extracting this. A standard textbook on expert systems explains the knowledge engineering bottleneck in exactly these sorts of terms: "The pieces of basic knowledge are assumed and combined so quickly that it is difficult for him (*the expert*) to describe the process" (Waterman 1986). Diana Forsythe in an anthropological study of knowledge engineers (Forsythe 1993) across a number of projects found that

> For knowledge engineers knowledge is apparently a 'thing' that can be extracted like a mineral or a diseased tooth.
>
> (p459)

And again:

> Knowledge engineers, then, seem to think of knowledge elicitation as a sort of surgical or engineering task, in which the substrate to be mined or operated upon is a human brain.
>
> (p459)

The process of knowledge acquisition has since been reconceptualised as the domain expert and knowledge engineer collaborating in creating a model. In a footnote in 2015 review of knowledge elicitation techniques Shadbolt and Smart (2015) comment,

> It should be pointed out that although early conceptualizations of knowledge elicitation cast the process as one of extracting or mining knowledge from the heads of experts, more recent conceptualizations view the process as a modelling exercise. The idea is that the knowledge elicitor and domain expert work together in order to create a model of an expert's knowledge. This model may reflect reality to a greater or lesser extent.

In this approach a central term is elicitation rather than extracting; however, creating "a model of the expert's knowledge" still suggests knowledge is some sort of stuff. We are not criticising the paper, simply pointing out that it is difficult to move away from the idea that knowledge is some sort of stuff, a set of concepts and logical relations, somehow independent of minds.

Despite this being a strong assumption about one of the most profound questions in philosophy, knowledge engineers will often shy away from any philosophical reflection on the nature of knowledge and intelligence. Forsythe notes the attitude of the knowledge engineers she interviewed:

> Rather than debate about the definition of 'intelligence', or discuss whether AI is or is not part of the computer science, they prefer to get on with their systems.
>
> (p457)

and quotes a lab member characterising the approach as:

> Lets not stop to think, lets just do it and we'll find out.
>
> (p457)

The idea or rather unreflected assumption that knowledge is something concrete probably arises from our use of language based on symbols. When we say a word, generally the person we are talking to will know what we are referring to or talking about. What does it mean that we use symbols? For Plato and Aristotle our use of symbols meant that we were getting to the form or essence of the things we were talking about. For example, there are 100s of different breeds of horse, from draught horses through to the dog-sized Falabella. To recognise such a wide range of animals as the same, they must share something in common. Plato's solution to this problem was that somewhere there must be a set of archetypes, including an archetypal concept of horse, and the horses we see in the world are some sort of faint shadow or reflection of the archetype. He used the story of the cave with a fire and with dancing shadows on the cave wall. The world around us is made up of shadows of the archetypes.

The influence of Plato seems clear in modern work on ontologies. There is no doubt that it is useful to have a set of terms to be used consistently – and even better to have a formal structure for the concepts and relationships the terms express. But modern work on ontologies also includes the development of so-called upper ontologies which cover ideas like object, quality, attribute, property, quantity, processes, events, and the ultimate goal is a shared universal ontology enabling reasoning across the Semantic Web. Despite the importance of developing shared standard ontologies, it is not at all clear whether they will be able to be used by humans. The Unified Medical Language System (UMLS) (not actually an ontology) has over one million concepts, five million synonyms and hundreds of terminologies. Rosenbloom et al. have pointed out that the most important issue with the UMLS is not its completeness, but how end-users can be enabled to use it (Rosenbloom et al. 2006). A recent study of some biomedical ontologies in use found significant errors (Boeker et al. 2011). Along with errors in ontologies, a further issue is their multiplicity so that ontology merging is a major research issue, but the challenge is not simply that different terminologies have emerged but that people disagree even on ontology alignment (Tordai et al. 2011). On the other hand, Tordai et al. comment: "Humans rarely have problems with disambiguating the meaning of words in a discourse context". In an earlier simpler example Shaw and Woodward carried out a study comparing different domain experts, as well as repeat studies with the same experts as a control (Shaw and Woodward 1988). They found experts construct knowledge with different terminologies and disagree with each other's terminologies but that this is not a problem as they are well used to both working together and disagreeing with each other!

This leads to the question of whether there is a more useful philosophy of knowledge than that knowledge is some sort of set of symbols and their relations. Winograd and Flores emphasised that a human expert always works in a context, every conversation is in a context and the expert responds to nuances in the context so that a computer program using symbols to reason independent of context is unlikely to be able to reproduce the subtlety of an expert (Winograd and Flores 1987). Clancey culminated a series of papers with his book on situated cognition[2] (Clancey 1997). In Clancey's view, knowledge does not exist in the head to be mined or modelled but is created and expressed in a particular context and for that context. This is why experts only ever provide partial knowledge. Situated cognition also provides an approach to understanding why knowledge acquisition is a collaborative modelling activity, not of what is in the expert's head, but of the world itself. Brian Gaines situates knowledge acquisition for knowledge-based systems as only the latest stage in a long history of mankind creating and systematising knowledge about the world (Gaines 2013).

Recognising the situated constructivist nature of knowledge gives a different framework for understanding knowledge acquisition for knowledge-based systems, but does not directly lead to practical techniques for knowledge acquisition. Another framework for understanding the situated nature of knowledge is Karl Popper's notion of falsification (Popper 1963). In Popper's view no hypothesis is ever proven; rather to support a new hypothesis we disprove the alternatives, generally the currently accepted wisdom. Popper applied this not only to scientific hypotheses, but to all hypotheses and beliefs. Kuhn and others have argued that science also evolves by other means than falsification (Kuhn 1962). For example, the systems biology approach to understanding biological systems widely used in biology and physiology in the 20th century was not disproven, rather the focus (and funding) changed simply because genetics and molecular biology seemed to be far more promising. But, Popper is clearly right when it comes to proving hypotheses – that a new hypothesis is not proven rather the alternative is disproven. This doesn't mean we know nothing, and everything is a guess. Lonergan's massive tome "Insight" (Lonergan 1959) is about the fundamental human capacity to discover and recognise the intelligibility in things. The classic example is Archimedes' "Eureka" when he discovered the principle of flotation. But discovering some intelligibility does not mean we have discovered and understood the thing-in-itself. Our insights, the intelligibility we have discovered, are always only a partial understanding. Because they are only partial, they are eventually falsified and we come up with new richer insights. Gaines' history of knowledge acquisition is essentially a history of discovering insight, then disproving and improving it. For those interested in the history of theology, the early Councils of the Church did exactly the same thing. They appear to be defining creeds and dogma, but on a closer look what they are really doing is rejecting hypotheses as not making sense and coming up with better hypotheses, expressed as dogmas. There are obviously deep philosophical questions in all of this, but there are also very practical implications for knowledge acquisition.

A little reflection suggests that disproving hypotheses is generally how we conduct everyday arguments. We rarely attempt to roll out some sort of systematic proof that we are right, rather we attempt to come up with the one-punch killer argument that will knock our opponent's position out of the ring (although it rarely does). It seems clear that the same sort of thing happens in expert judgement. A clinician, never formally proves a person has disease X, rather they exclude the other likely possibilities in the context – but there is always the possibility, no doubt extremely slight, that a person might have some other condition, that not all possible hypotheses have been considered and rejected. For example, apparently the following diseases all have similar if not identical symptoms in the early stages: Meningitis, Leptospirosis, Yellow fever, Haemorrhagic fever, Dengue fever, Malaria, Lyme disease, Rift valley fever, Legionnaire's disease, Hendra virus and Plague (and now the SARS-CoV-2 virus). Obviously, when a patient presents with these symptoms in Sweden, versus New Guinea versus Kenya, a clinician will have quite different sets of hypotheses they are trying to exclude in ordering diagnostic tests to come to a diagnosis. As another example, the Hendra virus occurs in Australia and a few other tropical countries. It is extremely rare with only seven reported cases in humans in Australia, four of which were fatal, as there is no treatment. Although extremely rare, are Australian clinicians more likely to consider this a remote possibility for a patient if there has been a news report of another Hendra fatality? Would Hendra ever be considered by a clinician in Sweden or Norway? These are of course exaggerated suppositions, and one can make the point that rule-based systems are trying to include relevant standard knowledge, not all possible knowledge. But this means that knowledge in a knowledge-based system is always incomplete, there is always the possibility of some other explanation for the data.

This modus operandi of excluding only other hypotheses likely in the context shows up most clearly when an expert explains why they reach one conclusion for a case than rather another. What they invariably do is point to features in the data that are consistent with conclusion X but which don't occur with cases where conclusion Y applies, or they point to features that occur with Y but not with X. Basing knowledge acquisition on the identification of discriminating features is not only exploiting obvious human ability but is also consistent with the situated cognition philosophy of knowledge outlined above. It is also fundamental to the notion of rationality. If a person wants to draw two different conclusions about two sets of data, they must have identified some difference in the data. Perhaps they are mistaken, but rationality requires that a different decision about two sets of data must relate to some apparent difference in the data, otherwise it is capricious and irrational. This requirement to identify differences includes techniques such as those based on Kelly's Personal Construct Psychology (PCP) (Shaw 1980, Gaines and Shaw 1993). In this approach a person is asked to think of three objects in the domain of interest and then asked to identify an attribute or construct whereby two of the objects are similar and dissimilar from the third. Repeatedly applying this approach gradually elicits a rich representation of the domain.

Ripple-Down Rules (RDR) are based on the same notion of differentiating cases but focus on asking the expert to identify features in the data for the cases which differentiate them. The PCP approach is aimed at creating a model for a domain, prior to or independent of any data representation of cases in the domain, whereas RDR generally assumes there is already a data representation and the role of the expert is to identify relevant features in the data representation of actual cases. These sorts of techniques simply bypass the question of whether or not experts have difficulty in reporting on how they make decisions. For example,

> Often the expertise has become so routinized that experts no longer know how they accomplish particular tasks.
>
> (Shadbolt and Smart 2015)

Techniques like RDR do not expect or ask the experts to explain how they accomplish particular tasks, but only to identify features that justify why conclusion X should be made about a case rather than Y. They first identify the features they believe lead to conclusion X, but if these apply to a case for which conclusion Y was previously assigned they are presented with a previous case and asked to identify some feature(s) that differentiate the cases. On the principle of rationality, if an expert says the X case is different from the Y case, even if they were mistaken in this, they must have identified a differentiating feature to be able to claim the case should be X rather than Y. The more likely problem with a weaker expert is that they might select less important differentiating features, which means that more cases will need to be seen in building the knowledge base.

NOTES

1 The ideas in this chapter were first presented in Compton and Jansen (1990).

2 For a brief history of situated cognition in knowledge acquisition, see (Compton 2013).

CHAPTER 3

Ripple-Down Rule Overview

This chapter outlines the key ways in which Ripple-Down Rules (RDR) differ from conventional knowledge acquisition methods. These features of RDR will be further clarified in the worked examples in the following chapters.

There are four key features of the RDR approach:

- Rules are only added to deal with specific cases.
- The order in which rules are evaluated is specified in the knowledge base rather than by the inference engine.
- Rules are only added, never removed or edited.
- Rule actions do not retract facts, but only assert facts.

3.1 CASE-DRIVEN KNOWLEDGE ACQUISITION

Users are never asked for a rule, as some sort of general statement, rather they are asked which features in a particular case led them to assign a particular conclusion for that case. A user might wish to assign a number of perhaps interdependent conclusions for a case, but they are asked about only one conclusion at a time, and the features that led them to that particular conclusion. In this discussion, a case is simply the set of data about which one or more decisions are to be made. For example, it might be the latest set of laboratory results for a patient, or it might be all the laboratory results available for the patient since investigations were commenced. It might also include age, sex, weight and clinical notes about the patient's medical history. In summary, a case is the data a user considers in reaching a conclusion about that data.

It may be that the set of features the user selects for the case in hand might also apply to some previous case where the RDR system assigned a different conclusion. If so, the user is asked to select one or more further features that distinguish the cases (or to agree that the conclusion for the previous case should be changed). Again, this is simply an application of the Principle of Rationality: if a user gives two different conclusions for two cases, they must have identified some feature or features that differentiate the cases. If there are no distinguishing features, it is not rational to claim that the cases are different. In practice there might be a number of such cases, with perhaps various conclusions, that should be distinguished from the current case. If so, the user is asked to select differentiating features, case by case, until the rule (the set of features selected) applies only to the new case, or the user has decided one of these past cases should have a different conclusion from its previous conclusion.

Generally with RDR, the previous cases against which the rule is checked are the specific cases for which rules have been added, which are called *cornerstone cases*. One could check a wider range of cases that have previously been processed, but as a minimum, cases which have promoted the addition of rule should not have their classification changed by a new rule, unless the user explicitly decides to change the conclusion for the earlier case.

We might also note that developing an ontology or appropriate representation for the domain is not the major concern for RDR as it often is in other knowledge engineering research. An implicit assumption with an RDR approach is that since the RDR system will deal with a stream of cases, the cases already have an appropriate representation.

3.2 ORDER OF CASES PROCESSED

The cases for which rules are added can be processed in any order; there is no requirement to add more general rules first. This enables RDR systems to be built while they are in use. Cases are processed by the evolving RDR system but the output is monitored; and whenever a case is detected as not having the correct conclusion, a rule is added. Pacific Knowledge Systems customers generally take only a few minutes to add a rule, with the median time being a minute or two (Compton et al. 2011, Compton 2013).

Rule redundancy can occur, but mainly if a user adds a rule that is very specific and closely matches the actual case resulting in rules being added unnecessarily. On the other hand, there is little problem in adding an over-general rule, as its scope will be rapidly reduced with the addition of refinement rules to correct errors from the over-generalisation. This does not mean the output of the RDR system has to be monitored permanently. All AI systems are (or should be) introduced into routine use only after appropriate performance on unseen data. A user monitoring the output of an RDR system can be provided with statistics on how often corrections have been required to enable them to make a judgement about whether the system still requires monitoring. Users can also set up different levels of on-going monitoring for different rules and different conclusions. Pacific Knowledge Systems has been using an approach like this for many years [1].

The order of cases and the quality of the rules added are much more likely to result in redundancy with Single Classification RDR (SCRDR). As will be seen, SCRDR results in a binary tree and so if a very general rule is added, but with irrelevant rule conditions, it will result in the same correction rules being added to both the true and false branches of the rule with the irrelevant conditions. In practice this does not happen as users invariably try to add rules with relevant conditions, and if they do add irrelevant conditions, it is generally for a rule that is over-precise, which is less likely to need refinement. The decision tree for the first large SCRDR system in routine use with 2000 rules was very unbalanced (see Figure 5.24); it was more a decision list with refinements also being decision lists (Preston, Edwards, and Compton 1994). This suggests that this potential problem was not a major issue with a large previously in-use SCRDR knowledge base.

3.3 LINKED PRODUCTION RULES

Expert or knowledge-based systems are invariably described as having two components: a production rule and inference engine. Typically (OMG 2009), a production rule is described as having two components: a condition part and an action list.

```
if [condition] then [action-list]
```

A knowledge-based system consists of a whole set of such rules, all independent and with no information about the order in which rules are evaluated. A very crude inference engine may simply evaluate rules in the order in which they are listed in a file (the knowledge base), but generally the inference engine will do something more sophisticated to decide which rule to evaluate next, known as a conflict resolution strategy. Conflict resolution strategies may include the following:

- Prefer rules with more conditions, i.e. more features in the case are matched by the rule.
- Prefer rules that use the most recently asserted fact in a rule condition.
- Pick the first rule in the knowledge base that satisfies the case (the crude strategy above).
- Alternatively, pick the last rule (the most recently added).
- Rules may also have extra information assigned, e.g. a numerical score known as salience. If two rules can fire, the one with the higher salience is chosen.
- If conflict resolution strategies don't result in one candidate rule, pick one at random.

The problem with an inference engine controlled by a conflict resolution strategy is that when a rule is added or changed it can be difficult to predict how other rules will behave, as was discussed in Chapter 1. This makes the case-differentiation strategy suggested above more difficult, because a case may no longer be evaluated by the same rules when other rules change.

In contrast RDR specify which rule is to be evaluated next. Most RDR knowledge bases end up as binary or n-ary trees, but a more general description is that RDR rules are *linked production rules:*

```
if [condition] then[case action-list],
                    [inference action]
else                [inference action]
```

With this representation the *case* action-list is the normal inference engine action asserting one or more conclusions or facts, or specifying some other actions. But there is also an inference action specified for whether the rule fires or fails to fire. In RDR as they have been developed so far, the *inference* action simply specifies what rule is to be evaluated next; that is, the path of a case through the knowledge base is completely determined. When a new rule is added it is linked to a previous rule using the *inference* action of the previous rule. So, the only previous cases which can reach this new rule are the cases which were evaluated by the previous rule with the same outcome thus allowing the RDR case-differentiation approach to work.

Information that can be used with standard rules such as salience can also give a rule a priority in determining whether a rule should be evaluated, but the order in which rules will be evaluated may change as new rules with different salience are added which the conflict resolution strategy takes into account. In our approach each rule contains information that determines which rule will be evaluated next, and once this new linked rule is added, the link information can't be changed. The closest to this work is Grosof's proposal for courteous logic, where a priority relationship between two rules that could fire is specified, giving them a kind of refinement relationship (Grosof 1997, 2004). More generally, Gabbay has proposed labelled logics able to specify priorities (Gabbay 1996). The OMG production rule specification also includes a specification for sequential processing as something distinct from inferencing where rules are simply evaluated in order (OMG 2009). RDR are processed sequentially, as in the OMG specification, but the order of processing is determined by where rules are added as refinements as well as the order in which they are added. In general, RDR uses a depth-first approach with older rules and their refinements evaluated before newer rules and their refinements. How this works in practice will be seen in the following chapters.

3.4 ADDING RULES

The conditions in the body of a rule cannot be changed, added to or deleted in the RDR approach, for the same reason as linked production rules are required. If a rule body can be edited, it becomes more challenging to predict what will happen with previous cases. Similarly, rules cannot be deleted. On the other hand, the conclusion of any rule can be changed, as this does not change the inference, only the label assigned to that inference path. However, such a step may have other implications depending on the application of the system. For example, if the label is used to specify a folder for a document management system, one might need to consider the impact on other documents already filed.

With RDR, rules are only added for a case either when a required conclusion has not been given or when the wrong conclusion has been given.

The rule added to correct a wrong conclusion is essentially a refinement rule for the rule giving the wrong conclusion and the conclusion that gets asserted by the refinement rule simply overrides the conclusion given by the parent rule. Normally a refinement rule reduces the scope of the parent rule, but a refinement rule could be identical to the old rule, but this is the same as simply changing the conclusion for the rule. A refinement can also be a "stopping" rule; that is, the rule has no conclusion, so no conclusion is given if the refinement rule fires. How this works in practice will be seen in the examples in the following chapters.

3.5 ASSERTIONS AND RETRACTIONS

Standard knowledge-based systems allow rules to assert facts (assign conclusions) and also to retract facts. In contrast RDR only allows facts to be asserted, except in the special case where a refinement rule replaces the conclusion of the rule that it is refining. Fact retraction is not required because a fact that would need to be retracted is a fact that has been incorrectly asserted. The RDR approach deals with this by adding a refinement rule to prevent the incorrect conclusion being given in the first place; that is, rather than using inference to manage conclusions that should not have been asserted, knowledge acquisition is used to prevent the incorrect conclusion being made in the first place. Again, how this works in practice will be seen in the following chapters.

3.6 FORMULAE IN CONCLUSIONS

In the discussion above, we assume conclusions are assertions of facts; however, similar to other rule-based system approaches a rule action can be a formula, a piece of code or basically anything that can be passed to and used by the world outside the rule-based system. Bekmann used RDR to determine the formulae for the mutation operator and fitness function for a genetic algorithm program, resulting in different formulae being used in different circumstances as the generations progressed (Bekmann and Hoffmann 2005, Bekmann and Hoffmann 2004). Misra developed an RDR-based system to determine which processes, or which parameters for a process, should be used in complex multi-process tasks such as image processing (Misra, Sowmya, and Compton 2010). Drake and Beydoun have also proposed predicate-logic-based RDR (Drake and Beydoun 2000).

In Chapter 7 we will further distinguish between conclusions which are assertions of fact and conclusions which are actually outputs from the RDR system specifying some sort of action on the outside world. As will be discussed, a rule which specifies an action on the world and does not have a refinement rule attached terminates inference as control is passed back to the system that has called the RDR.

NOTE

1 https://pks.com.au/wp-content/uploads/2015/03/WhitePaperAutovalidationRippleDownPKS.pdf

CHAPTER 4

Introduction to Excel_RDR

The following chapters provide worked examples of different types of Ripple-Down Rules (RDR). The software used for these examples are a series of demonstrator RDR systems written in Excel and VBA. The reason for using Excel is that it allows data presentation and manipulation that is very widely used, so providing RDR for this environment should make it possible for a wide audience to play with the technology.

Excel_RDR is available at http://www.cse.unsw.edu.au/~compton/RDR_software.html. It will also become available on general download sites under the name Excel_RDR. A detailed manual is available with the download, but the following should be sufficient to get started.

Although the Excel_RDR systems are full featured, they are not designed for routine use in knowledge-based applications in industry. The purpose of the Excel-RDR systems is to make the RDR algorithms and methods sufficiently clear, so that a developer can readily build their own RDR system in the environment best suited to integrating RDR with the target information system for a project. Excel_RDR could be used for an industrial project, but it will likely be too slow for large-scale batch processing and interfacing will be clumsy (and the code is probably too buggy). One could also export a compiled knowledge base which could be converted to run with some other language rather than VBA, but again in the long term it is better to write your own RDR.

There is a detailed manual available with the download, but if you bypass the manual to go direct to the programs note that you must first trust access to the VBA object model, through the Trust Centre under

Options with Windows Excel or through the Security setting under Preferences with Mac Excel. The reason for this requirement is to allow Excel_RDR to write its own VBA code to speed up inference, but this makes your machine very vulnerable, so make sure you turn off trusted access before you import any Excel files from a non-trusted source.

Any data can be used with Excel_RDR, but in the following demos we use the simple Zoo dataset from the UCIrvine data repository[1], as anyone should be able to at least guess rules for this dataset and it is a well-known standard dataset. This dataset is already loaded into the Excel_RDR spreadsheets and a portion of the dataset is shown below in Figure 4.1. In the UCIrvine dataset TRUE is represented as 1 and FALSE as 0, so the same representation is used here. The Zoo dataset included with Excel_RDR is exactly as downloaded from UCIrvine. There will be later discussion about the quality of datasets for machine learning as it turns out even this tiny dataset has some errors.

This sheet is for Cases you will evaluate with the RDR KB
Enter attribute names in the next row in place of attribute 1, 2 etc, and then Cases below

name	*hair*	*feathers*	*eggs*	*milk*	*airborne*	*aquatic*	*predator*	*toothed*	*backbone*	*breathes*	*venomous*	*fins*	*no of legs*	*tail*	*domestic*	*catsize*	*target*	*conclusion*
aardvark	1	0	0	1	0	0	1	1	1	1	0	0	4	0	0	big	mammal	
antelope	1	0	0	1	0	0	0	1	1	1	0	0	4	1	0	big	mammal	
bass	0	0	1	0	0	1	1	1	1	0	0	1	0	1	0	small	fish	
bear	1	0	0	1	0	0	1	1	1	1	0	0	4	0	0	big	mammal	
boar	1	0	0	1	0	0	1	1	1	1	0	0	4	1	0	big	mammal	
buffalo	1	0	0	1	0	0	0	1	1	1	0	0	4	1	0	big	mammal	
calf	1	0	0	1	0	0	0	1	1	1	0	0	4	1	1	big	mammal	
carp	0	0	1	0	0	1	0	1	1	0	0	1	0	1	1	small	fish	
catfish	0	0	1	0	0	1	1	1	1	0	0	1	0	1	0	small	fish	
cavy	1	0	0	1	0	0	0	1	1	1	0	0	4	0	1	small	mammal	
cheetah	1	0	0	1	0	0	1	1	1	1	0	0	4	1	0	big	mammal	
chicken	0	1	1	0	1	0	0	0	1	1	0	0	2	1	1	small	bird	
chub	0	0	1	0	0	1	1	1	1	0	0	1	0	1	0	small	fish	
clam	0	0	1	0	0	0	1	0	0	0	0	0	0	0	0	small	mollusc	
crab	0	0	1	0	0	1	1	0	0	0	0	0	4	0	0	small	mollusc	
crayfish	0	0	1	0	0	1	1	0	0	0	0	0	6	0	0	small	mollusc	
crow	0	1	1	0	1	0	1	0	1	1	0	0	2	1	0	small	bird	
deer	1	0	0	1	0	0	0	1	1	1	0	0	4	1	0	big	mammal	

FIGURE 4.1 Part of the Zoo dataset shown on the Excel_RDR data input worksheet.

To make the screenshots more readable, we have deleted columns that were not required when we created rules by hand and also were not used in the machine learning experiments on this data (Figure 4.2).

If you double-click on any case (a row), the menu in Figure 4.3 is shown. The first button takes you to the screen where you select rule conditions to build a rule for that case. The second button shows a rule trace for that case. The next three buttons run a series of cases. *Run till error* assumes the correct classification is in the second-last column labelled *target,* which is how UCIrvine datasets are normally set up. For the other *run* commands, any row where the *target* differs from the *conclusion* assigned by rules (in the

This sheet is for Cases you will evaluate with the RDR KB													
Enter attribute names in the next row in place of attribute 1, 2 etc, and then Cases below													
name	*hair*	*feathers*	*eggs*	*milk*	*airborne*	*aquatic*	*backbone*	*breathes*	*fins*	*no of legs*	*tail*	*target*	*conclusion*
aardvark	1	0	0	1	0	0	1	1	0	4	0	mammal	
antelope	1	0	0	1	0	0	1	1	0	4	1	mammal	
bass	0	0	1	0	0	1	1	0	1	0	1	fish	
bear	1	0	0	1	0	0	1	1	0	4	0	mammal	
boar	1	0	0	1	0	0	1	1	0	4	1	mammal	
buffalo	1	0	0	1	0	0	1	1	0	4	1	mammal	
calf	1	0	0	1	0	0	1	1	0	4	1	mammal	
carp	0	0	1	0	0	1	1	0	1	0	1	fish	
catfish	0	0	1	0	0	1	1	0	1	0	1	fish	
cavy	1	0	0	1	0	0	1	1	0	4	0	mammal	
cheetah	1	0	0	1	0	0	1	1	0	4	1	mammal	
chicken	0	1	1	0	0	0	1	1	0	2	1	bird	
chub	0	0	1	0	0	1	1	0	1	0	1	fish	
clam	0	0	1	0	0	0	0	0	0	0	0	mollusc	
crab	0	0	1	0	0	1	0	0	0	4	0	mollusc	
crayfish	0	0	1	0	0	1	0	0	0	6	0	mollusc	
crow	0	1	1	0	0	0	1	1	0	2	1	bird	
deer	1	0	0	1	0	0	1	1	0	4	1	mammal	

FIGURE 4.2 Part of the Zoo dataset with irrelevant columns removed.

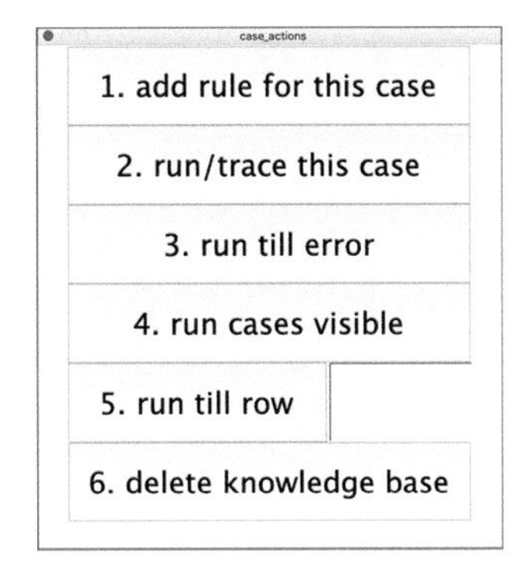

FIGURE 4.3 Excel_RDR control palette.

conclusion column) is coloured red. The final command deletes the knowledge base but leaves the cases intact for repeat experiments.

If the *add rule for this case* button is clicked, the system goes to the rule builder worksheet where rule conditions can be specified as follows:

In the example in Figure 4.4, a fruitbat had been misclassified as a *bird* whereas it should have been classified as a *mammal.* The *bird* rule that fired incorrectly for the fruitbat was originally added for a chicken, so both the fruitbat case and the chicken cornerstone case are shown. Any features in the cornerstone case which were used in reaching the conclusion are coloured orange. The user builds a rule in the section labelled *new rule* and selects rule conditions in the form *attribute name – operator*

	This sheet is where you build a rule for a case						
	name	*hair*	*feathers*	eggs	milk	airborne	aquatic
cornerstone	chicken	0	1	1	0	1	0
current case	fruitbat	1	0	0	1	1	0
	name	*hair*	*feathers*	*eggs*	*milk*	*airborne*	*aquatic*
new rule	*operator*	*operator*	*operator*	*operator*	*is*	*operator*	*operator*
	value	*value*	*value*	*value*	1	*value*	*value*

FIGURE 4.4 A portion of the rule builder worksheet showing the selection of a rule condition.

– *value.* In the example in Figure 4.4, the operator *is* and the value 1 are selected so the rule condition is *milk is 1.* Because this is a valid condition for the fruitbat case, the rule condition and the milk feature in the fruitbat case are shown as green, but since this condition is false for the chicken cornerstone case, the milk feature for that case is coloured pink. The user can select any number of further conditions to add to the rule. If the rule will fire on the cornerstone case, the user is asked whether they really want the cornerstone conclusion changed or not.

The built-in operators available for creating rule conditions in Excel_RDR are as follows:

- the numerical operators >, < , >=, <=, =,
- the case-independent Boolean/text operator "is",
- the case-independent text operator "contains".

Users can also create their own operators. The VBA module *user_funcs* already contains examples of user-defined functions defined in VBA, such as *both, before, just_before, is_missing, is_something, is_not.* These can be added to or replaced as needed to deal with a particular dataset. The manual contains more details about how to do this. The system will warn the user if an operator call is invalid.

Excel_RDR has other features such as the statistics sheet and functions for running cornerstone cases. These are not covered here because they are not essential to understanding RDR, but they are described in the manual accompanying the software; however, most of what you can do with Excel_RDR should become obvious in the following worked examples.

Excel_RDR also allows the user to specify that the conclusion is any VBA function taking input parameters from the case. Examples of this will be shown in Chapter 7 where string functions combining other conclusions into a suitable output are used in the final conclusion.

NOTE

1 https://archive.ics.uci.edu/ml/datasets/Zoo

CHAPTER 5

Single Classification Example

Single Classification RDR (SCRDR) allows only a single conclusion for a case. For example, in the Zoo data base, each animal can belong to only one category, such as mammal, fish, bird etc. If only a single classification is possible, more complex classifications such as flying mammal, aquatic mammal, land mammal, flying bird, aquatic bird, land bird, result in a lot of individual classifications. A more efficient approach would be multiple classifications, e.g. an animal would be assigned two classifications, such as mammal and flying, but in this chapter, we are considering a system that gives only a single classification. Standard machine learning systems such as J48 are single classification learners. Chapter 6 will consider an RDR system that gives multiple independent conclusions for a case.

Knowledge acquisition event: The first case in the data base in Figure 5.1 is the aardvark. If the user double clicks on the aardvark row and then selects *1. Add rule for this case* from the pop-up menu (Figure 4.3), the system goes to the rule builder worksheet as shown in Figure 5.1.

The user then selects one or more rule conditions and a conclusion. Since no rule had fired on the aardvark case, there is no cornerstone case (Figure 5.1). The user selects *milk = 1* as a rule condition and enters conclusion *mammal* as shown in Figure 5.2. It seems reasonable that this single condition is enough to define a mammal.

This sheet is where you build a rule for a case

	name	*hair*	*feathers*	eggs	milk	airborne	aquatic	back bone	breathes	fins	no of legs	tail	target	old conclusions
cornerstone														
current case	aardvark	1	0	0	1	0	0	1	1	0	4	0	mammal	
	name	*hair*	*feathers*	*eggs*	*milk*	*airborne*	*aquatic*	*backbone*	*breathes*	*fins*	*no of legs*	*tail*	*target*	*conclusion*
new rule	*operator*	*operator*	*operator*	*operator*	*operator*	*operator*	*operator*	*operator*	*operator*	*operator*	*operator*	*operator*	*operator*	enter new conclusion
	value	*value*	*value*	*value*	*value*	*value*	*value*	*value*	*value*	*value*	*value*	*value*	*value*	here

add this rule to KB

FIGURE 5.1 The aardvark case shown on the rule builder worksheet.

This sheet is where you build a rule for a case

	name	*hair*	*feathers*	eggs	milk	airborne	aquatic	back bone	breathes	fins	no of legs	tail	target	old conclusions
cornerstone														
current case	aardvark	1	0	0	1	0	0	1	1	0	4	0	mammal	
	name	*hair*	*feathers*	*eggs*	*milk*	*airborne*	*aquatic*	*backbone*	*breathes*	*fins*	*no of legs*	*tail*	*target*	*conclusion*
new rule	*operator*	*operator*	*operator*	*operator*	=	*operator*	*operator*	*operator*	*operator*	*operator*	*operator*	*operator*	*operator*	mammal
	value	*value*	*value*	*value*	1	*value*	*value*	*value*	*value*	*value*	*value*	*value*	*value*	

add this rule to KB

FIGURE 5.2 A rule condition and conclusion for the aardvark case.

If the user then clicks the *add this rule to KB* button shown towards the bottom of Figure 5.2, the screen in Figure 5.3 is shown, where the rule is displayed in conventional IF.... THEN fashion and the user is asked to confirm that they want to add the rule.

	name	*hair*	*feathers*	eggs	milk	airborne	aquatic	backbone	breathes	fins	no of legs	tail	target	old conclusions
cornerstone														
current case	aardvark	1	0	0	1	0	0	1	1	0	4	0	mammal	
	name	*hair*	*feathers*	*eggs*	*milk*	*airborne*	*aquatic*	*backbone*	*breathes*	*fins*	*no of legs*	*tail*	*target*	*conclusion*
new rule	*operator*	*operator*	*operator*	*operator*	=	*operator*	*op*					*ator*	*operator*	mammal
	value	*value*	*value*	*value*	1	*value*	*va*						*value*	

Microsoft Excel

Are you sure you want to add this rule to the KB

No Yes

IF milk = 1 THEN conclusion = mammal

add this rule to KB

FIGURE 5.3 The rule confirmation dialogue box.

If the rule is added, the user is returned to the case worksheet and the case for which the rule was added is shown with the new conclusion and the row is coloured green. If the user then selects *run till error*, the following screen is shown (Figure 5.4).

This sheet is for Cases you will evaluate with the RDR KB													
Enter attribute names in the next row in place of attribute 1, 2 etc, and then Cases below													
name	***hair***	***feathers***	***eggs***	***milk***	***airborne***	***aquatic***	***backbone***	***breathes***	***fins***	***no of legs***	***tail***	***target***	***conclusion***
aardvark	1	0	0	1	0	0	1	1	0	4	0	mammal	mammal
antelope	1	0	0	1	0	0	1	1	0	4	1	mammal	mammal
bass	0	0	1	0	0	1	1	0	1	0	1	fish	

FIGURE 5.4 The result from selecting *run till error* after the rule in Figure 5.3 has been added.

Knowledge acquisition event: For the first and second rows, the conclusion correctly matched the target, but as there is no rule yet for fish, for the third case no conclusion was given and *run till error* inference stopped. We then added a rule for this case as shown in Figure 5.5.

	name	*hair*	*feathers*	eggs	milk	airborne	aquatic	backbone	breathes	fins	no ofl egs	tail	target	old conclusions
cornerstone														
current case	bass	0	0	1	0	0	1	1	0	1	0	1	fish	
	name	*hair*	*feathers*	*eggs*	*milk*	*airborne*	*aquatic*	*backbone*	*breathes*	*fins*	*no of legs*	*tail*	*target*	*conclusion*
new rule	*operator*	*operator*	*operator*	*operator*	*operator*	*operator*	=	*operator*	*operator*	*operator*	*operator*	*operator*	*operator*	**fish**
	value	*value*	*value*	*value*	*value*	*value*	*1*	*value*	*value*	*value*	*value*	*value*	*value*	
	IF aquatic = 1 THEN conclusion = fish													

FIGURE 5.5 An overgeneralised rule for fish.

Clearly, this is an overgeneralised rule because there are many aquatic animals that are not fish. Although a domain expert would be very unlikely to give such an overgeneralised rule, it is used here to demonstrate what happens when overgeneralised rules are added. After adding the rule, we process cases until the next error, which occurs because a chicken is not classified. The rule shown in Figure 5.6 is then added.

	name	*hair*	*feathers*	eggs	milk	airborne	aquatic	backbone	breathes	fins	no of legs	tail	target	old conclusions
cornerstone														
current case	chicken	0	1	1	0	0	0	1	1	0	2	1	bird	
	name	*hair*	*feathers*	*eggs*	*milk*	*airborne*	*aquatic*	*backbone*	*breathes*	*fins*	*no of legs*	*tail*	*target*	*conclusion*
new rule	*operator*	*operator*	*is*	*operator*	*operator*	*operator*	*operator*	*operator*	*operator*	*operator*	*operator*	*operator*	*operator*	**bird**
	value	*value*	*1*	*value*	*value*	*value*	*value*	*value*	*value*	*value*	*value*	*value*	*value*	
	IF feathers is 1 THEN conclusion = bird													

FIGURE 5.6 A rule added to classify a chicken as a bird.

To our knowledge all birds have feathers, although it may turn out with a later case that there are featherless birds. It is one of the key features of RDR, that although you should try to add a useful rule, there is no need to try to think ahead and anticipate all the possible cases covered by the rule – you will find out soon enough.

Knowledge acquisition event: The next error is that a clam should be classified as a *mollusc*. We don't know how to define a *mollusc*, but the conditions in the rule shown in Figure 5.7 seemed to us to cover the key distinguishing characteristics of a clam. Again, you don't have to come up with perfect rule – if the rule is too general, you will eventually add a refinement rule, or if too narrow you will eventually add an extra rule. You do this if and when a case occurs for which the rule gives the wrong conclusion.

	name	*hair*	*feathers*	eggs	milk	airborne	aquatic	backbone	breathes	fins	no of legs	tail	target	old conclusions
cornerstone														
current case	clam	0	0	1	0	0	0	0	0	0	0	0	mollusc	
	name	*hair*	*feathers*	*eggs*	*milk*	*airborne*	*aquatic*	*backbone*	*breathes*	*fins*	*no of legs*	*tail*	*target*	*conclusion*
new rule	*operator*	*operator*	*operator*	*operator*	*operator*	*operator*	*operator*	=	=	*operator*	=	*operator*	*operator*	mollusc
	value	*value*	*value*	*value*	*value*	*value*	*value*	*0*	*0*	*value*	*0*	*value*	*value*	

IF backbone = 0 AND breathes = 0 AND no of legs = 0 THEN conclusion = mollusc

FIGURE 5.7 A guessed rule for a mollusc.

The rules added so far can be seen on the rules worksheet below. To get to this sheet, we double-clicked on the third row of the case sheet, showing the data for a bass and then selected *run/trace this case* from the pop-up menu.

The row at the top of Figure 5.8 shows the case we chose to trace when on the case worksheet. If this row on the rules worksheet is edited and *return* is clicked, the modified case is rerun. Or if any of the rule rows below are double-clicked then the cornerstone case for that rule (the case for which the rule was created) is placed in the case row. The next four rows are the four rules added so far and show the rule trace. Red means a row has been evaluated but the rule fails to fire for the case, green means the rule did fire for the case, and orange is used to indicate a rule that was not evaluated in this case, because evaluation stopped after the conclusion *fish* was reached.

This sheet contains the rules you have built; you can edit the case, but not the rules

				name	*hair*	*feathers*	*eggs*	*milk*	*airborne*	*aquatic*	*backbone*	*breathes*	*fins*	*no of legs*	*tail*	*target*	*conclusion*
			case	bass	0	0	1	0	0	1	1	0	1	0	1	fish	fish
order added	Go to if true	Go to if false	Rule no	*name*	*hair*	*feathers*	*eggs*	*milk*	*airborne*	*aquatic*	*backbone*	*breathes*	*fins*	*no of legs*	*tail*	*target*	*conclusion*
1	exit	2	1					= 1									mammal
2	exit	3	2							= 1							fish
3	exit	4	3			is 1											bird
4	exit	exit	4								= 0	= 0		= 0			mollusc

FIGURE 5.8 The first four rules added.

The "inference actions" for these linked production rules are shown on the left of Figure 5.8. If the first rule had fired the system would have exited with the conclusion from the first rule. In this case it doesn't fire so the inference action is *Go to if false*. Inference therefore goes to rule 2 and that rule is evaluated. For this case the second rule fires on the case and the conclusion *fish* is given. Since the second rule fired, the inference action *Go to if true* results in system exiting and rules 3 and 4 are not evaluated. So far this is obviously a simple decision list with each new rule added to the bottom. The first column shows the order in which the rules were added, while on the other side of the two "inference action" columns the rule number is given. Rules are numbered depth first. Since no refinement rules have been added they are the same so far.

Knowledge acquisition event: Run till error is selected again and inference stops when a crab is misclassified as a *fish*, because as anticipated the single condition *aquatic = 1* for *fish* was too general. The rule shown in Figure 5.9 is then added.

	name	*hair*	*feathers*	eggs	milk	airborne	aquatic	backbone	breathes	fins	no of legs	tail	target	old conclusions
cornerstone	bass	0	0	1	0	0	1	1	0	1	0	1	fish	fish
current case	crab	0	0	1	0	0	1	0	0	0	4	0	mollusc	fish
	name	*hair*	*feathers*	*eggs*	*milk*	*airborne*	*aquatic*	*backbone*	*breathes*	*fins*	*no of legs*	*tail*	*target*	*conclusion*
new rule	*operator*	*operator*	*operator*	*operator*	*operator*	*operator*	*operator*	=	*operator*	=	*operator*	*operator*	*operator*	mollusc
	value	*value*	*value*	*value*	*value*	*value*	*value*	0	*value*	0	*value*	*value*	*value*	

IF backbone = 0 AND fins = 0 THEN conclusion = mollusc

FIGURE 5.9 A rule to classify a crab as a mollusc.

Again we are not sure how to define a *mollusc*, but in Figure 5.9 we are shown the cornerstone case for which the fish rule was added. Our task is simply to select features that we think apply to *molluscs* like a crab – with at least one of them not applying to the bass cornerstone case. This means that the rule we add will correctly classify a crab as a *mollusc*, but will not classify a *fish* like a bass as a *mollusc*. In this guess we select two features that we think clearly distinguish a crab from a fish. Again, the key feature of the RDR approach is that the user does not need to spend time developing the perfect rule – although of course they should enter the best rule they can. There is no point in taking a lot of time to try to make a perfect rule since, as we have already discussed, there is no such thing as the perfect rule. As long as the rule applies to the new case, but not to the cornerstone case, the rule is adequate but may be refined further if a later case requires this. It probably should be noted at this stage that although RDR rules can't be edited or deleted, a user can change the conclusion given by a rule. The same rule applies in the same circumstance, so it covers the same concept, but the conclusion or concept has a different name. In Excel_RDR this is done simply by editing the conclusion cell for the rule on the rules worksheet. We can also note that the user might decide that the new rule should apply to the cornerstone case. If the rule will fire on the cornerstone case, they are asked to confirm that this is what they want.

After the rule shown in Figure 5.9 is added the rules are as shown in Figure 5.10 showing the rule trace for a crab.

				name	*hair*	*feathers*	*eggs*	*milk*	*airborne*	*aquatic*	*backbone*	*breathes*	*fins*	*no of legs*	*tail*	*target*	*conclusion*
			case	crab	0	0	1	0	0	1	0	0	0	4	0	mollusc	mollusc
order added	**Go to if true**	**Go to if false**	**Rule no**	*name*	*hair*	*feathers*	*eggs*	*milk*	*airborne*	*aquatic*	*backbone*	*breathes*	*fins*	*no of legs*	*tail*	*target*	*conclusion*
1	exit	2	1					= 1									**mammal**
2	3	4	2							= 1							**fish**
5	exit	exit	3								= 0		= 0				**mollusc**
3	exit	5	4			is 1											**bird**
4	exit	exit	5								= 0	= 0		= 0			**mollusc**

FIGURE 5.10 The first refinement rule added.

It can be seen in Figure 5.10 that the new rule for *mollusc* is added as a refinement rule under the rule for *fish* that classified the crab as a *fish* inappropriately. The inference action panel on the left shows that inference goes from rule 2 to rule 3 if rule 2 fires. This contrasts with the rules before the *mollusc* rule was added in Figure 5.8 where if rule 2 fires, the system exits with *fish* from rule 2 as the conclusion. RDR manages this rule placement automatically; the user need not pay any attention to the structure that is emerging. As can be seen from the *order added* and the *rule no* columns, rule no 3 was actually the 5th rule added.

The structure that is now starting to emerge is a simple binary tree (Figure 5.11)

```
IF milk is 1 THEN conclusion is mammal
ELSE IF aquatic = 1 THEN conclusion is fish
EXCEPT IF backbone = 0 AND fins = 0 THENconclusion is mollusc
ELSE IF feathers is 1 THEN conclusion is bird
ELSE IF backbone = 0 AND breathes = 0 and no of legs = 0 THEN conclusion is mollusc
END IF
```

FIGURE 5.11 shows Figure 5.10 as rules in a binary tree.

Knowledge acquisition event: The next error found in continuing to process cases is that a duck is misclassified as a *fish* and the rule below in Figure 5.12 is added.

	name	hair	feathers	eggs	milk	airborne	aquatic	backbone	breathes	fins	no of legs	tail	target	old conclusions
cornerstone	bass	0	0	1	0	0	1	1	0	1	0	1	fish	fish
currentcase	duck	0	1	1	0	0	1	1	1	0	2	1	bird	fish
	name	*hair*	*feathers*	*eggs*	*milk*	*airborne*	*aquatic*	*backbone*	*breathes*	*fins*	*no of legs*	*tail*	*target*	*conclusion*
new rule	*operator*	*operator*	=	*operator*	*operator*	*operator*	*operator*	*operator*	*operator*	*operator*	*operator*	*operator*	*operator*	**bird**
	value	*value*	*1*	*value*	*value*	*value*	*value*	*value*	*value*	*value*	*value*	*value*	*value*	
	IF feathers = 1 THEN conclusion = bird													

FIGURE 5.12 A rule to change the conclusion from *fish* to *bird* for a duck.

As anticipated above, *aquatic* was not a sufficiently precise condition for classifying a *fish*, so the rule has to be refined to enable the *bird* conclusion. This also shows a potential problem with SCRDR. We had already added an identical rule:

IF feathers THEN conclusion = bird

This was the third rule added, but the second rule was the rule for *fish* using the condition *aquatic = 1*, so this is the rule that fired and has to be refined. This suggests the same rule may appear many times in the tree. Although potentially the same rule may appear in more than one place, it will not appear in many places unless it is an extreme overgeneralisation, which is unlikely if a person with some expertise about the domain develops the knowledge base. As noted above, we were well aware our rule for *fish* was too general.

In the rules so far, shown in Figure 5.13, the rule added to classify a duck as a *bird* rather than a *fish* is the second refinement for rule 2, and so is only evaluated if the first refinement for *mollusc* fails, shown by the rule trace in Figure 5.13; that is, the system fires rule 6 and gives conclusion *bird* for a duck because these conditions are satisfied:

1. NOT (Milk = 1)
2. aquatic = 1
3. NOT (backbone = 0 and legs = 0)
4. feathers = 1

As can be seen, even though we have only six very simple rules, the knowledge being used to reach a conclusion is starting to become a little bit complex. But this is not of concern to the user; the user keeps simply adding either a rule to correct a conclusion, or a rule to give a conclusion where there wasn't one before.

				name	hair	feathers	eggs	milk	airborne	aquatic	backbone	breathes	fins	no of legs	tail	target	conclusion
			case	duck	0	1	1	0	0	1	1	1	0	2	1	bird	bird
order added	Go to if true	Go to if false	Rule no	name	hair	feathers	eggs	milk	airborne	aquatic	backbone	breathes	fins	no of legs	tail	target	conclusion
1	exit	2	1					= 1									mammal
2	3	5	2							= 1							fish
5	exit	4	3								= 0		= 0				mollusc
6	exit	exit	4			= 1											bird
3	exit	6	5			is 1											bird
4	exit	exit	6								= 0	= 0		= 0			mollusc

FIGURE 5.13 The rules, after the rule in Figure 5.12 is added to change the conclusion for a duck from *fish* to *bird*.

Knowledge acquisition event: The next error is that a flea is not classified. We add the rule shown in Figure 5.14 as another guess and since no rule had fired for the flea, the new rule is automatically added at the end of the rules.

	name	hair	feathers	eggs	milk	airborne	aquatic	backbone	breathes	fins	no of legs	tail	target	old conclusions
cornerstone														
current case	flea	0	0	1	0	0	0	0	1	0	6	0	insect	
	name	*hair*	*feathers*	eggs	*milk*	*airborne*	*aquatic*	backbone	breathes	*fins*	*no of legs*	*tail*	*target*	*conclusion*
new rule	*operator*	*operator*	*operator*	=	*operator*	*operator*	*operator*	=	=	*operator*	*operator*	*operator*	*operator*	**insect**
	value	*value*	*value*	1	*value*	*value*	*value*	0	1	*value*	*value*	*value*	*value*	

IF eggs = 1 AND backbone = 0 AND breathes = 1 THEN conclusion = insect

FIGURE 5.14 A rule for an *insect*; no rule previously fired.

Knowledge acquisition event: The next error is yet again because our overgeneralised rule that anything that is aquatic (*aquatic = 1*) is a *fish* has now misclassified a frog as a *fish* rather than an *amphibian*. We add the condition *breathes = 1*, but this time trying to be smarter, we realise that there are other aquatic creatures such a dolphin or platypus that are also aquatic and breathe, so we also add the condition that they don't secrete milk (*milk = 0*) as shown in Figure 5.15.

	name	hair	feathers	eggs	milk	airborne	aquatic	backbone	breathes	fins	no of legs	tail	target	old conclusions
cornerstone	bass	0	0	1	0	0	1	1	0	1	0	1	fish	fish
current case	frog	0	0	1	0	0	1	1	1	0	4	0	amphibian	fish
	name	*hair*	*feathers*	*eggs*	milk	*airborne*	*aquatic*	*backbone*	breathes	*fins*	*no of legs*	*tail*	*target*	*conclusion*
new rule	*operator*	*operator*	*operator*	*operator*	=	*operator*	*operator*	*operator*	=	*operator*	*operator*	*operator*	*operator*	amphibian
	value	*value*	*value*	*value*	0	*value*	*value*	*value*	1	*value*	*value*	*value*	*value*	

IF milk = 0 AND breathes = 1 THEN conclusion = amphibian

FIGURE 5.15 A rule to classify a frog as an *amphibian* rather than a *fish*.

The *milk = 0* condition is redundant because the very first rule has the condition *milk = 1*, so a *mammal* that is aquatic would have been captured by this rule and classified as a *mammal*, with inference stopping. However, the key value of RDR is that the user does not need to think about what other rules might fire on some other cases – and this gets very challenging if there are thousands of rules. The whole point of RDR is that users only have to worry about the case in front of them and the cornerstone case that resulted in the rule which gave the wrong conclusion. The resultant knowledge base after the rule in Figure 5.15 is automatically added is shown in Figure 5.16. The rule trace for the frog case as shown in Figure 5.16 is:

- rule 2 the *fish* rule fires
- rules 3 and 4, the earlier refinement rules for rule 2 are evaluated but fail to fire
- rule 5, the third refinement rule, fires correcting the *fish* conclusion to *amphibian*

				name	*hair*	*feathers*	*eggs*	*milk*	*airborne*	*aquatic*	*backbone*	*breathes*	*fins*	*no of legs*	*tail*	*target*	*conclusion*
			case	frog	0	0	1	0	0	1	1	1	0	4	0	amphibian	
order added	**Go to if true**	**Go to if false**	**Rule no**	*name*	*hair*	*feathers*	*eggs*	*milk*	*airborne*	*aquatic*	*backbone*	*breathes*	*fins*	*no of legs*	*tail*	*target*	*conclusion*
1	exit	2	1					= 1									**mammal**
2	3	6	2							= 1							**fish**
5	exit	4	3								= 0		= 0				**mollusc**
6	exit	5	4			= 1											**bird**
8	exit	exit	5					= 0				= 1					**amphibian**
3	exit	7	6			is 1											**bird**
4	exit	8	7								= 0	= 0		= 0			**mollusc**
7	exit	exit	8				= 1				= 0	= 1					**insect**

FIGURE 5.16 Rule trace for a frog, after rule 8 (Figure 5.15) was added.

Knowledge acquisition event: The next error was for a pitviper, and since this was the first *reptile* seen, no rule fired so the system adds the new rule at the bottom. The rule we add is shown in Figure 5.17. Again, we were not sure how to define a *reptile*, but this rule seems to represent a reasonable guess.

Knowledge acquisition event: The next error is that a scorpion is not classified, when it should be classified as a *mollusc*, so again the rule is automatically placed at the end. As before we are not really sure how to define a dry land *mollusc*. The rule we add is shown in Figure 5.18.

	name	*hair*	*feathers*	eggs	milk	airborne	aquatic	back bone	breathes	fins	no of legs	tail	target	old conclusions
cornerstone														
current case	pitviper	0	0	1	0	0	0	1	1	0	0	1	reptile	
	name	*hair*	*feathers*	*eggs*	*milk*	*airborne*	*aquatic*	*backbone*	*breathes*	*fins*	*no of legs*	*tail*	*target*	*conclusion*
new rule	*operator*	*operator*	*operator*	*operator*	*operator*	*operator*	*operator*	=	=	*operator*	=	=	*operator*	reptile
	value	*value*	*value*	*value*	*value*	*value*	*value*	*1*	*1*	*value*	*0*	*1*	*value*	
	IF backbone = 1 AND breathes = 1 AND no of legs = 0 AND tail = 1 THEN conclusion = reptile													

FIGURE 5.17 A *reptile* rule for a pitviper.

	name	*hair*	*feathers*	eggs	milk	airborne	aquatic	backbone	breathes	fins	no of legs	tail	target	old conclusions
cornerstone														
current case	scorpion	0	0	0	0	0	0	0	1	0	8	1	mollusc	
	name	*hair*	*feathers*	*eggs*	*milk*	*airborne*	*aquatic*	*backbone*	*breathes*	*fins*	*no of legs*	*tail*	*target*	*conclusion*
new rule	*operator*	*operator*	*operator*	*operator*	*operator*	*operator*	*operator*	=	=	*operator*	>=	*operator*	*operator*	mollusc
	value	*value*	*value*	*value*	*value*	*value*	*value*	*0*	*1*	*value*	*4*	*value*	*value*	
	IF backbone = 0 AND breathes = 1 AND no of legs > = 4 THEN conclusion = mollusc													

FIGURE 5.18 A *mollusc* rule for a scorpion.

Knowledge acquisition event: The next error we find is that a seasnake is classified as *fish* rather than a *reptile* because of our earlier *fish* rule with the condition *aquatic = 1*. We add the rule shown in Figure 5.19.

	name	*hair*	*feathers*	eggs	milk	airborne	aquatic	backbone	breathes	fins	no of legs	tail	target	old conclusions
cornerstone	bass	0	0	1	0	0	1	1	0	1	0	1	fish	fish
current case	seasnake	0	0	0	0	0	1	1	0	0	0	1	reptile	fish
	name	*hair*	*feathers*	*eggs*	*milk*	*airborne*	*aquatic*	*backbone*	*breathes*	*fins*	*no of legs*	*tail*	*target*	*conclusion*
new rule	*operator*	*operator*	*operator*	*operator*	*operator*	*operator*	*operator*	=	*operator*	=	*operator*	*operator*	*operator*	reptile
	value	*value*	*value*	*value*	*value*	*value*	*value*	*1*	*value*	*0*	*value*	*value*	*value*	
	IF backbone = 1 AND fins = 0 THEN conclusion = reptile													

FIGURE 5.19 A *reptile* rule added to correct a seasnake misclassified as a *fish*.

This rule is again a guess, and perhaps we could have added *no of legs = 0* as an obvious feature of a snake, but instead we tried a more general rule that it has a backbone (*backbone = 1*) and no fins (*fins = 0*). We needed to add either *fins = 1* or *eggs = 0* or both, as these are the only features that distinguish a snake from the *fish* cornerstone case. We didn't use *eggs = 0* as we thought that reptiles generally did lay eggs. We didn't need *backbone = 1*, as it wasn't a discriminating condition, but added it because we had been confused about *molluscs,* but were pretty sure *molluscs* never have backbones.

Knowledge acquisition event: The next error is that a slug is misclassified as an *insect* rather than a *mollusc* – and again we have to make a guess as to how to define a *mollusc*. However, it seems pretty obvious that a major difference between an *insect* and slug is that a slug has no legs, so we create the following rule (Figure 5.20).

	name	hair	feathers	eggs	milk	airborne	aquatic	backbone	breathes	fins	no of legs	tail	target	old conclusions
cornerstone	flea	0	0	1	0	0	0	0	1	0	6	0	insect	insect
current case	slug	0	0	1	0	0	0	0	1	0	0	0	mollusc	insect
	name	*hair*	*feathers*	*eggs*	*milk*	*airborne*	*aquatic*	*backbone*	*breathes*	*fins*	*no of legs*	*tail*	*target*	*conclusion*
new rule	*operator*		*operator*	*operator*	*operator*	*operator*	*operator*	*operator*	*operator*	*operator*	=	*operator*	*operator*	mollusc
	value	*value*	*value*	*value*	*value*	*value*	*value*	*value*	*value*	*value*	0	*value*	*value*	

IF no of legs = 0 THEN conclusion = mollusc

FIGURE 5.20 A rule added to correct the misclassification of a slug as an *insect* to a *mollusc*.

Knowledge acquisition event: The next error is that a tortoise is not classified. Again, we guess at what might be the distinguishing characteristics that make it a *reptile*. It was tempting to add *legs >= 4*, as an obvious characteristic of a tortoise, but we chose to leave the rule more general. The rule as written would also apply to *birds*, but no feathered creature will ever reach this rule as they will be picked up by the third rule added. We make this observation because this book is concerned with explaining RDR, but in normal use a user would not need to think about what rules might have been added previously or how inference proceeds (Figure 5.21).

	name	hair	feathers	eggs	milk	airborne	aquatic	backbone	breathes	fins	no of legs	tail	target	old conclusions
cornerstone														
currentcase	tortoise	0	0	1	0	0	0	1	1	0	4	1	reptile	
	name	*hair*	*feathers*	*eggs*	*milk*	*airborne*	*aquatic*	*backbone*	*breathes*	*fins*	*no of legs*	*tail*	*target*	*conclusion*
new rule	*operator*	*operator*	*operator*	=	*operator*	*operator*	*operator*	=	=	*operator*	*operator*	=	*operator*	reptile
	value	*value*	*value*	1	*value*	*value*	*value*	1	1	*value*	*value*	1	*value*	

IF eggs = 1 AND backbone = 1 AND breathes = 1 AND tail = 1 THEN conclusion = reptile

FIGURE 5.21 Rule to classify a tortoise as a *reptile*.

No further errors occurred with the remaining cases, and the full knowledge base with 13 rules is shown in Figure 5.22. The rule trace for the tortoise, the final case for which a rule was added is shown. That rule was the only rule evaluated as true; seven rules were evaluated but failed, and five rules were not evaluated. The rules that were not evaluated were of course refinements, which were not reached because their parent rules, the overgeneralised rule for a fish and the rule for an insect, did not fire.

With this knowledge base, if no rule fires for a case then no conclusion is given. Alternatively, a default conclusion could be given, such as

				name	hair	feathers	eggs	milk	airborne	aquatic	backbone	breathes	fins	no of legs	tail	target	conclus ion
			case	tortoise	0	0	1	0	0	0	1	1	0	4	1	reptile	reptile
order added	Go to if true	Go to if false	Rule no	name	hair	feathers	eggs	milk	airborne	aquatic	backbone	breathes	fins	no of legs	tail	target	conclusion
1	exit	2	1					= 1									mammal
2	3	7	2							= 1							fish
5	exit	4	3								= 0		= 0				mollusc
6	exit	5	4			= 1											bird
8	exit	6	5					= 0				= 1					amphibian
11	exit	exit	6								= 1		= 0				reptile
3	exit	8	7			is 1											bird
4	exit	9	8								= 0	= 0		= 0			mollusc
7	10	11	9				= 1				= 0	= 1					insect
12	exit	exit	10											= 0			mollusc
9	exit	12	11								= 1	= 1		= 0	= 1		reptile
10	exit	13	12								= 0	= 1		>= 4			mollusc
13	exit	exit	13				= 1				= 1	= 1			= 1		reptile

FIGURE 5.22 The complete knowledge base.

"unknown animal". This is done simply by entering a rule on the rule builder sheet that doesn't have a body; this is described in the manual which is part of the Excel_RDR software package.

5.1 REPETITION IN AN SCRDR KNOWLEDGE BASE

As noted above, the obvious issue for SCRDR knowledge bases is the potential for repetition if rules using irrelevant conditions are added – particularly if added early. For example, in a medical diagnosis knowledge base, if the patient's gender was irrelevant to the diagnoses being made, there would likely be significant repetition if this rule:

```
IF sex = Male THEN disease X
```

was added as the first rule. The knowledge base would probably be double in size as the rules with actually relevant conditions would have to be included under both the true and false branch of this first rule.

If the knowledge base in Figure 5.22 is represented as a binary decision tree as in Figure 5.23, we see there is only one level of correction in this simple knowledge base, with only two rules being corrected. The four corrections for rule 2 were because of the overgeneralised rule for *fish*:

```
IF aquatic THEN Fish
```

As noted above, this was a deliberate overgeneralisation to demonstrate how SCRDR deals with overgeneralisation. If a better rule had been chosen for *fish*, we would have needed only one rule for *bird* rather than two. Likewise, we may have needed only two rules for *reptile*, rather than three. For an *amphibian, aquatic = 1* is an essential attribute. This suggests that with a better rule for *fish*, we would have needed two–three rather than

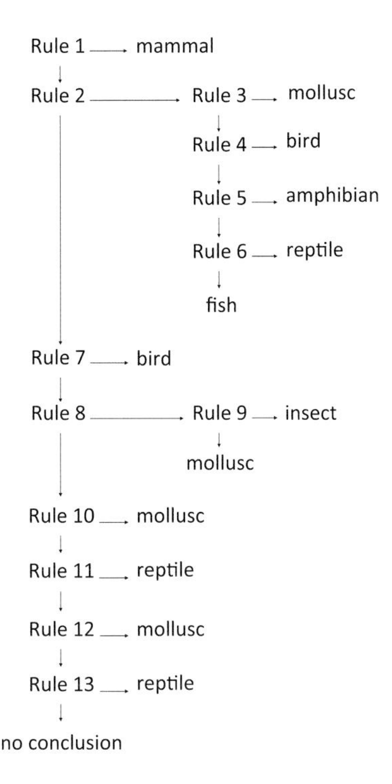

FIGURE 5.23 Figure 5.22 as a binary tree.

four rules. We assume in any serious application the rule conditions a user selects to build the knowledge base will be able to do considerably better than selecting random or peripherally relevant conditions.

In the first large industrial SCRDR system in production use, the only repetition problem noted was because of the restriction to single classifications. This meant that combinations of conclusions had to be expressed as a single classification, resulting in the same components of a rule having to appear in other rules as well (Edwards et al. 1993). This knowledge base eventually grew to about 2,000 rules but there were no complaints from the pathologist writing rules about repetition, apart from the multiple classification issue (Preston, Edwards, and Compton 1994). The decision tree from this knowledge base is shown in Figure 5.24, but presented sideways unlike Figure 5.23. Repetition in this knowledge base was not measured, and probably could not be. In Figure 5.23, overgeneralised rule 2 results in a number of corrections, so there is gap along the spine between rule 2 and rule 7. If overgeneralisation was a major problem for the large knowledge base in Figure 5.24, we might similarly expect to see some significant gaps along the spine. This does not seem to show up in Figure

5.24, although we cannot draw any definitive conclusion from this. But we note again that the knowledge base builder did not complain about the method resulting in repeatedly having to add the same rule (except for combination conclusions). If we assume the knowledge builder has a reasonable level of expertise, we believe the risk of repeatedly having to add the same rule is small. Simulation studies with varying levels of simulated expertise also suggest that repetition is not a major problem (Compton, Preston, and Kang 1995, Cao and Compton 2005, 2006a, b). On the other hand, Multiple Classification RDR, discussed in Chapter 6, removes this risk – at the cost of a small extra effort during knowledge acquisition. Cao et al. also showed formally that an SCRDR knowledge base must converge, regardless of the quality of the rules (Cao, Martin, and Compton 2004). In practice in commercial MCRDR systems it takes only a few minutes to add a rule (Compton et al. 2006, Compton et al. 2011). A detailed example of an in-use knowledge base shows that the median time to add a rule for a 3,000 rule knowledge base was a minute or two (Compton 2013). Since SCRDR (with at most one cornerstone case) requires less effort than MCRDR, rule addition is a very minor task, so one would not expert users to notice repetition unless for some reason it was a major problem.

FIGURE 5.24 A 2000 rule SCRDR decision tree from (Preston, Edwards, and Compton 1994).

We have built a small knowledge base for the Zoo data, but since we processed all the training data to build these 13 rules, there is no evidence that it will work on unseen cases (except the last few after tortoise, which did not require a further rule). Secondly, as repeatedly noted above, we (the authors) only have an everyman understanding of the taxonomy of the animal kingdom. We should also note that this means every time we built an Zoo knowledge base, we tended to use slightly different rules as we tried to make alternative guesses and the rules guessed would then determine which further guesses needed to be made. One would expect then that machine learning might do much better than guessing rules.

5.2 SCRDR EVALUATION AND MACHINE LEARNING COMPARISON

To evaluate the SCRDR approach we used 10-fold cross-validation, the standard evaluation approach in machine learning; that is, we manually built an SCRDR knowledge base using only 90% of the cases and tested it on the remaining 10% of cases. We did this 10 times with a different 10% test set each time. We also randomised the order of the 90% training cases each time, as obviously the order in which cases are seen may affect the resultant knowledge base. We also recorded the time taken to develop each of the 10 knowledge bases. Only one of us was involved in building the 10 knowledge bases. For comparison we also carried out a 10-fold cross-validation study using machine learning. We used J48 from the Weka machine learning workbench (Holmes, Donkin, and Witten 1994). J48, based on C4.5 (Quinlan 1992), produces a binary tree with each split based on an attribute value. Since all the attributes in the Zoo dataset apart from *no of legs* are Boolean, this is a very simple learning task. The attribute value selected for a split by J48 is the value that gives the best separation of the classes overall. There can be a number of ways of assessing this with the original being information gain. The contrast between the J48 tree and the SCRDR binary tree is that with SCRDR, the split is not on a single attribute value, but on the conjunction of conditions in a rule. [1]

The results are shown with and without pruning (Table 5.1). Without pruning was used just in case pruning was a problem with the small number of cases, but as usually happens, there were better results with pruning using the J48 default – although the difference here is negligible.

TABLE 5.1 The number of correct and error cases for the Zoo dataset with SCRDR and machine learning

Method	correct cases	error cases
J48 pruned	93	8
J48 unpruned	92	9
Manual SCRDR	99	2

Perhaps surprisingly SCRDR works better than J48. However, perhaps this is not so surprising when one recalls that J48 is based on purely statistical measures, so if the training data does not contain a sufficient number of the right examples, it will fail to learn. A human user, even when guessing, is making guesses based on a lifetime of experience that may be of some use, even in a domain they know little about. In Chapter 1 we discussed the IBM project on Indian street addresses where RDR rules based on human experience significantly outperformed machine learning methods on data that was perhaps somewhat different from the training data (Dani et al. 2010)

Although the SCRDR knowledge base for the Zoo domain outperformed J48, there were two cases that were not correctly classified in the SCRDR 10-fold cross-validation studies. The two cases were the scorpion and the seasnake. These two cases would have appeared in the 10% test data only once each, as there is a different 10% test data for each of the 10 folds. This suggests that these cases were such that unless the user actually saw them, rules would not be written that would cover them – i.e. they are unlike any other cases in the dataset. The reason they are unlike any other case is because in fact there are errors in the data. Wikipedia suggests that a scorpion, rather than being *mollusc*, is an *arachnid*, making it pure luck that the features a user might select to identify an actual *mollusc* also apply to a scorpion. It is also the only *arachnid* in the dataset. The error with the seasnake data is that all seasnakes breathe whereas according to the Zoo dataset a seasnake does not breathe. There is also a question of whether *eggs = 0* is appropriate for a seasnake. Seasnakes do have eggs, but apparently their eggs hatch internally rather than being laid before they hatch. We also noticed that exactly the same case for a frog appeared twice in the dataset.

We removed the scorpion case and corrected the seasnake data so *breathes = 1* and redid the J48 machine learning (Table 5.2). There was virtually no change to the J48 results, but removing the scorpion data removed one SCRDR error. In the fold where the seasnake data ended up in the test data, the *reptile* rule that could have applied, had *eggs = 1* as a condition, so depending how one understands the *eggs* feature, there were either no errors or one error for the user-developed SCRDR.

TABLE 5.2 The number of correct and error cases for a corrected Zoo dataset with SCRDR and machine learning

Method	correct cases	error cases
J48 pruned	92	8
J48 unpruned	93	7
Manual SCRDR	99–100	0–1

Two errors out of 100[2] cases raise other questions about the quality of the dataset. If a case is very unusual it will tend to be used for building rules in many of the folds and the cases used most frequently for building rules across the folds are shown in Table 5.3.

TABLE 5.3 The number of times various cases required rules across the 10 folds

name	target	No. of times rules were added for this case
scorpion	*Mollusc/arachnid*	9
slug	*mollusc*	8
frog	*amphibian*	7
tuatara	*reptile*	5
crab	*mollusc*	4
newt	*amphibian*	4
seasnake	*reptile*	4

The scorpion is the most unique in that every time it occurred in the training data a specific rule had to be added, but a slug is almost as unique. With examples of *molluscs* occurring three times in this table, clearly *molluscs* contain a variety of different sub-species with not much in common. The reason the frog appears more often is because it occurs twice in the four instances of *amphibian* in the dataset as downloaded from UCIrvine. *Reptiles* occur only 5 times in the dataset, but *reptiles* with legs are very different from reptiles without legs.

The other relevant data on the performance of SCRDR is how long the SCRDR development took and the size of the knowledge bases compared to the machine learning knowledge bases. J48 produced 8-9 rules and learning was of course almost instantaneous. For SCRDR there was an average of 10.6 rules over the 10 folds with the maximum being 14 rules. The average time to build a rule was 38 secs with the maximum being 69 secs. Such quick rule-building is to be expected given that for large real-world knowledge bases the median rule building time is a minute or two (Compton 2013). We initially assumed that the extra RDR rules compared to J48 were the reason for the better performance. This does not seem to be the case, as in repeating the 10-fold cross-validation study to produce the graph in Chapter 8, the rule builder tried to build better rules than the overgeneralised rules in this chapter – to behave more like an expert. The average knowledge base size reduced to 8.9 rules, similar to J48, with a maximum of 10 rules, and with the same two error cases. And interestingly, despite his efforts to come up with good rules, and experience in rebuilding the same knowledge base again and again, in every knowledge base there were still rules that required refinement!

We are not suggesting for a moment that RDR should be used instead of machine learning if there is a large well-curated dataset with reliable labelling of the data. For the Zoo dataset, we assume that machine learning did poorly compared to SCRDR because of the small number of examples; no doubt a much larger dataset would give results as good as SCRDR. However, what is of interest is the errors in this small dataset. There is no suggestion in the UCIrvine documentation that there were deliberate errors in the dataset. And although, one would assume these errors must have been noticed, we cannot find any references to these errors being discovered in machine learning studies using this data; however, we have only done a search, not examined in detail the over 5,000 papers that refer to this dataset. We have noted in Chapter 1 that the provision of large, accurate and well-curated datasets is one of the major issues in using machine learning to build knowledge bases capable of subtle decision making. What we have seen here is that there are errors even in a very tiny standard dataset.

5.3 SUMMARY

As the worked example above demonstrates, SCRDR greatly simplify knowledge acquisition such that a user with minimal training (and limited expertise) can reliably build a knowledge base that compares more than favourably with machine learning.

SCRDR is not a general-purpose knowledge-based-system technology; similar to most supervised machine learning methods it produces a knowledge base that classifies each data case as belonging to one of a number of mutually exclusive classes. For more complex tasks one of the other RDR methods should be used or a number of SCRDR knowledge bases should be used together (e.g. Dani et al. 2010)

Finally, there is a risk with the SCRDR binary tree structure of having to add the same rules repeatedly. The risk is that a rule may be added based on irrelevant attributes so that the same more relevant knowledge will have to be added under both the true and false branch of such rules. It is difficult to imagine an organization would assign a knowledge building task to someone who was going to make up rules based on irrelevant attributes. Of course, an expert rule builder may (and will) overgeneralise, but this doesn't necessarily mean significant redundancy, it just means the over-general rule will have to be refined. On the other hand, there is considerable potential for repetition if each single classification is a combination of other classifications. This problem is handled by Multiple Classification RDR described in Chapter 6.

NOTES

1 There are various machine learners based on the SCRDR representation, as discussed in Appendix 2.
2 There were 101 cases in the dataset but the frog case was duplicated.

CHAPTER 6

Multiple Classification Example

6.1 INTRODUCTION TO MULTIPLE CLASSIFICATION RIPPLE-DOWN RULES (MCRDR)

As the name suggests, MCRDR provides possibly many conclusions for a case, with the conclusions being independent. With SCRDR, once a rule fires, only refinements of that rule are evaluated. In contrast with MCRDR all rules without a parent are evaluated, and if a rule fires, all its refinements are evaluated. The structure that this leads to and how it compares to SCRDR will be discussed after the example is developed.

MCRDR also requires the possibility of "stopping rules", where a stopping rule is a refinement rule with no conclusion; that is, when a combination of independent conclusions is given for a case a user might decide that one particular conclusion should not be given. A stopping rule is the same as the previous EXCEPT IF rule in Figure 5.11 except that no conclusion is given:

```
IF A and B THEN conclusion = X
     EXCEPT IF C and D THEN conclusion = null
```

A rule can have many such stopping rules attached, and if any of the stopping rules fire the conclusion is not given. McCreath et al. have called rules with one or more stopping rules attached, composite rules (McCreath, Kay, and Crawford 2006, Crawford, Kay, and McCreath 2002). Kim et al.

(Kim, Compton, and Kang 2012) have noted that RDR rules with stopping rules could also be considered as censored production rules (Michalski and Winston 1986).

With MCRDR, if a rule gives a wrong conclusion, then a refinement rule can be added to change the conclusion in the same way as SCRDR. Alternatively, a stopping rule can be added so that the wrong conclusion is not given, plus a new rule to give the right conclusion.

If we are to be consistent in saying that MCRDR conclusions are independent, then probably refinement rules should not be allowed, as implicitly if one conclusion is replaced by another because of a refinement rule, then they are probably not independent but mutually exclusive. On the other hand, in real world applications such as interpreting chemical pathology data it is not clear cut that correcting one conclusion with another necessarily implies that they are mutually exclusive. However, refinement rules do not have to be used; one can always add both a stopping rule and a new rule to give the right conclusion. The stopping rule plus new rule has another advantage or difference in that all the rules are at the top level and will be evaluated for all cases (and a rule at the top level is simply a rule that will always be evaluated), whereas with refinement rules the parent rule has to fire for the refinement rules to be evaluated. If a rule is added at the top level, the cornerstone cases for all the previous rules can reach the new rule and need to be checked. If refinement rules are used then the cornerstone cases to be checked are the case from the parent rule and the cases from all the sibling refinements of that parent, as these can all reach the new refinement.

In the example here the stopping rule plus new rule strategy is used, rather than the refinement strategy. The Zoo domain is a single classification domain but MCRDR can handle single classification, simply by adding stopping rules to ensure only one conclusion is given. Later in this chapter we add a further conclusion about where the animal lives to demonstrate actual multiple conclusions.

6.2 EXCEL_MCRDR EXAMPLE

In this chapter we build an MCRDR knowledge base for the same Zoo domain. As far as possible we make the same overgeneralisations as for the SCRDR Zoo knowledge base, so a direct comparison is possible.

Knowledge acquisition event: As with SCRDR, we double-click the first aardvark case and select *add this rule to KB* from the pop-up menu and are taken to a slightly different screen from SCRDR as shown in Figure 6.1.

This sheet is where you build a rule for a case

		name	hair	feathers	eggs	milk	airborne	aquatic	backbone	breathes	fins	no of legs	tail	target	conclusion
run case	cornerstone														
	current case	aardvark	1	0	0	1	0	0	1	1	0	4	0	mammal	

	name	hair	feathers	eggs	milk	airborne	aquatic	backbone	breathes	fins	no of legs	tail	target	conclusion
new rule	operator	operator	operator	operator	operator	operator	operator	operator	operator	operator	operator	operator	operator	enter new conclusion here
	value	value	value	value	value	value	value	value	value	value	value	value	value	

add this rule to KB

FIGURE 6.1 The MCRDR rule builder screen.

Since MCRDR allows for multiple conclusions the *run case* button on the left gives one conclusion at a time, so the user can correct whichever of the conclusions is incorrect. Each time *run case* is clicked, the inference proceeds to the next rule that fires, and the conclusion for that rule is shown with a red border. When the *rule builder* screen is first shown for a case, there is no border round the conclusion field and the user must run the case before a rule can be selected.

When the *run case* button is clicked for an aardvark (Figure 6.1), the screen in Figure 6.2 is shown.

This sheet is where you build a rule for a case

		name	hair	feathers	eggs	milk	airborne	aquatic	backbone	breathes	fins	no of legs	tail	target	conclusion	
run case	cornerstone															
	current case	aardvark	1	0	0	1	0	0	1	1	0	4	0	mammal		final output:

	name	hair	feathers	eggs	milk	airborne	aquatic	backbone	breathes	fins	no of legs	tail	target	conclusion
new rule	operator	operator	operator	operator	operator	operator	operator	operator	operator	operator	operator	operator	operator	enter new conclusion here
	value	value	value	value	value	value	value	value	value	value	value	value	value	

add this rule to KB

FIGURE 6.2 The rule builder screen after run case button is clicked in Figure 6.1.

The red border round an empty conclusion field means no rule fired on that inference path and no conclusion was given. Note also there is an extra cell next to the conclusion containing *final output:* If a rule fires, the conclusion of that rule will go in the red conclusion box, and that conclusion will be added to the conclusions already in the *final output* cell, so the user can see where they are up to in terms of conclusions being added to a case. For the aardvark case a rule is then added, the same rule as in SCRDR (Figure 5.2). However, instead of returning to the *cases* worksheet as with SCRDR, the system goes to the *rules* worksheet (Figure 6.3). With Excel_MCRDR, inference is carried out automatically when the system goes to the *rules* worksheet where the rule trace for the case is shown.

This sheet contains the rules you have built; you can edit the case, but not the rules

				name	hair	feathers	eggs	milk	airborne	aquatic	backbone	breathes	fins	no of legs	tail	target	conclusion
			case	aardvark	1	0	0	1	0	0	1	1	0	4	0	mammal	mammal
order added	Go to if true	Go to if false	Rule no	name	hair	feathers	eggs	milk	airborne	aquatic	backbone	breathes	fins	no of legs	tail	target	conclusion
1	exit	exit	1					= 1									mammal

FIGURE 6.3 The rules worksheet after the first rule is entered.

We then continue testing cases and adding rules as required. The target column for this UCIrvine dataset contains only a single conclusion, but with MCRDR many conclusions may be given for a case, concatenated in the conclusion cell for the case on the *cases* worksheet. In this demonstrator, the *run till error* menu choice simply checks if any one of the conclusions is the target conclusion. Obviously manual inspection is also needed to check if the other conclusions are appropriate.

Knowledge acquisition event: We next add the same rule as SCRDR to classify a bass as a *fish*. This time when the *add this rule to KB* button is clicked, the system warns the user that it will check if the proposed rule will fire on any of the cornerstone cases as shown in Figure 6.4.

		name	hair	feathers	eggs	milk	airborne	aquatic	backbone	breathes	fins	no of legs	tail	target	conclusion	
run case	cornerstone															
	current case	bass	0	0	1	0	0	1	1	0	1	0	1	fish		final output:
		name	hair	feathers	eggs	milk	airborne	aquatic	backbone	breathes	fins	no of legs	tail	target	conclusion	
	new rule	operator	operator	operator	operator	operator	operator	=	operator	operator	operator	operator	operator	operator	fish	
		value	value	value	value	value	value	1	v							

Microsoft Excel

Cornerstones will now be tested to add the new rule giving this conclusion

OK

add this rule to KB

FIGURE 6.4 When a rule is added, cornerstone cases are first checked.

The same dialogue box did not appear for the first rule, because there were no previous rules, so no cornerstone cases. Once the first rule was added, there was a cornerstone case which might cause the new rule to fire. This is simply because we are adding rules at the top level, so every cornerstone case for a previous rule would now also reach the new rule, and perhaps cause it to fire. In this case the previous aardvark case does not fire the *fish* rule, so the system goes to the rules worksheet. Examples where cornerstone cases fire a proposed rule will be seen below.

Knowledge acquisition event: The third and fourth rules for a chicken as a *bird* and a clam as a *mollusc* are added in the same way, with the conditions in the four rules identical to SCRDR, and although cornerstone cases from the previous rules are checked, none fire the new rules. Figure 6.5

shows the MCRDR inference path for a bass which is quite different from the SCRDR inference path in Figure 5.8. When an SCRDR rule fires, only its refinement rules are evaluated, and with no refinement rules yet added to the SCRDR knowledge base in Figure 5.8 the *go to if true* for the first four rules is *exit*. With SCRDR, for the case of the bass, the first rule fails, the second fires and therefore the third and fourth are not evaluated (Figure 5.8). For MCRDR, for all four rules, the *go to if true* inference action goes to the next rule and all three rules are evaluated, but only the second rule fires. *Exit* is not reached until all the rules have been evaluated.

				name	hair	feathers	eggs	milk	airborne	aquatic	backbone	breathes	fins	no of legs	tail	target	conclusion
			c case	bass	0	0	1	0	0	1	1	0	1	0	1	fish	fish
order added	Go to if true	Go to if false	Rule no	name	hair	feathers	eggs	milk	airborne	aquatic	backbone	breathes	fins	no of legs	tail	target	conclusion
1	2	2	1					= 1									mammal
2	3	3	2							= 1							fish
3	4	4	3			is 1											bird
4	exit	exit	4								= 0	= 0		= 0			mollusc

FIGURE 6.5 The first four MCRDR rules (compare with the same first four SCRDR rules in Figure 5.8).

Knowledge acquisition event: As with the SCRDR knowledge base, the next error is that a crab is classified as a *fish* rather than a *mollusc*. When we click *run case* on the rule builder worksheet, we get the screen shown in Figure 6.6. This shows the crab being given the *fish* conclusion, but also the cornerstone case for the *fish* rule.

		name	hair	feathers	eggs	milk	airborne	aquatic	backbone	breathes	fins	no of legs	tail	target	conclusion	
run case		bass	0	0	1	0	0	1	1	0	1	0	1	fish		final output: fish
	current case	crab	0	0	1	0	0	1	0	0	0	4	0	mollusc	fish	*final output: fish*
		name	hair	feathers	eggs	milk	airborne	aquatic	backbone	breathes	fins	no of legs	tail	target	conclusion	
	new rule	operator	operator	operator	operator	operator	operator	operator	operator	operator	operator	operator	operator	operator	enter new conclusion here	
		value	value	value	value	value	value	value	value	value	value	value	value	value		

FIGURE 6.6 The result of clicking run case on the rule builder worksheet.

We construct a rule with the conditions: *backbone = 0* and *breathes = 0* as used in the SCRDR rules for this case and add the *mollusc* conclusion. However, in this implementation the rule is first automatically added as a stopping rule, that is without a conclusion, and the user is asked if this is OK as shown in Figure 6.7.

After it has added the stopping rule, the system moves to automatically add a new rule at the end of the list of rules to give the conclusion *mollusc*. The new rule starts with the two conditions from the stopping rule and the

		name	hair	feathers	eggs	milk	airborne	aquatic	backbone	breathes	fins	no of legs	tail	target	conclusion	
run case	cornerstone	bass	0	0	1	0	0	1	1	0	1	0	1	fish		final output: fish
	current case	crab	0	0	1	0	0	1	0	0	0	4	0	mollusc	fish	*final output: fish*

new ru

Microsoft Excel

this rule will first be entered as stopping rule added to prevent conclusion "fish". Is that OK?

No Yes

ıquatic	backbone	breathes	fins	no of legs	tail	target	conclusion
›perator	=	=	operator	operator	operator	operator	mollusc
›alue	0	0	value	value	value	value	

add this rule to KB

FIGURE 6.7 A stopping rule being added so that a crab is not classified as a *fish*.

user is advised that this rule will be then tested against cornerstone cases as shown in Figure 6.8 to see if any extra conditions need to be added to exclude any other cornerstone case. There are four cornerstone cases to be checked because if one follows the inference actions shown in Figure 6.5, the four cases that prompted the addition of these rules will all reach the new fifth rule being added. All the cornerstone cases always reach a new rule.

		name	hair	feathers	eggs	milk	airborne	aquatic	backbone	breathes	fins	no of legs	tail	target	conclusion	
run case	cornerstone	bass	0	0	1	0	0	1	1	0	1	0	1	fish		final output: fish
	current case	crab	0	0	1	0	0	1	0	0	0	4	0	mollusc	fish	*final output: fish*

new rul

Microsoft Excel

a stopping rule has been added; cornerstones will now be tested to add the new rule giving this conclusion

OK

aquatic	backbone	breathes	fins	no of legs	tail	target	conclusion
operator	=	=	operator	operator	operator	operator	mollusc
value	0	0	value	value	value	value	

add this rule to KB

FIGURE 6.8 The user being advised that the cornerstones will be tested.

The only cornerstone case that fires the rule is from the previous *mollusc* rule added for a clam. Since this has the same conclusion as the new rule, it is not shown to the user.

Figure 6.9 shows the resulting knowledge base with the rule trace for the crab data. The *fish* rule now has a stopping rule attached so if the *fish* rule is true, inference passes to rule 3, a stopping rule. The new rule was added at the bottom of the rules. The rule trace for a crab in Figure 6.9 shows the

			case	name	hair	feathers	eggs	milk	airborne	aquatic	backbone	breathes	fins	no of legs	tail	target	conclusion
				crab	0	0	1	0	0	1	0	0	0	4	0	mollusc	mollusc
order added	**Go to if true**	**Go to if false**	**Rule no**	***name***	***hair***	***feathers***	***eggs***	***milk***	***airborne***	***aquatic***	***backbone***	***breathes***	***fins***	***no of legs***	***tail***	***target***	***conclusion***
1	2	2	1					= 1									mammal
2	3	4	2							= 1							fish
5	4	4	3								= 0	= 0					
3	5	5	4			is 1											bird
4	6	6	5								= 0	= 0		= 0			mollusc
6	exit	exit	6								= 0	= 0					mollusc

FIGURE 6.9 The knowledge base with the rule trace for a crab

fish rule fires, but the conclusion is not given, because the stopping rule also fires. The final *mollusc* rule then fires giving the required conclusion.

Figure 6.10 shows the rule trace for the clam cornerstone case, which fired the new *mollusc* rule but was not shown to the expert, because the new rule gave the correct conclusion for the case – for the second time. The consequence of both rules firing is that *mollusc* appears twice in the conclusion cell for the case. In any real world application, the mollusc conclusion would only be shown once in the conclusion field, but this would mask the issues that arise with the same conclusion being given multiple times, so for this demonstrator all instances of the conclusion are shown. This is discussed further below.

				name	*hair*	*feathers*	*eggs*	*milk*	*airborne*	*aquatic*	*backbone*	*breathes*	*fins*	*no of legs*	*tail*	*target*	*conclusion*
			c case	clam	0	0	1	0	0	0	0	0	0	0	0	mollusc	mollusc. mollusc

order added	Go to if true	Go to if false	Rule no	*name*	*hair*	*feathers*	*eggs*	*milk*	*airborne*	*aquatic*	*backbone*	*breathes*	*fins*	*no of legs*	*tail*	*target*	*conclusion*
1	2	2	1					= 1									mammal
2	3	4	2							= 1							fish
5	4	4	3								= 0	= 0					
3	5	5	4			is 1											bird
4	6	6	5								= 0	= 0		= 0			mollusc
6	exit	exit	6								= 0	= 0					mollusc

FIGURE 6.10 The rule trace showing the two *mollusc* rules that fire for the clam cornerstone case.

Knowledge acquisition event: The next rule added is because a dolphin is classified as a *fish* as well as a *mammal*. It was classified as a *fish* because rule 2 fired, but the stopping rule, rule 3, did not fire as shown in Figure 6.11.

				name	*hair*	*feathers*	*eggs*	*milk*	*airborne*	*aquatic*	*backbone*	*breathes*	*fins*	*no of legs*	*tail*	*target*	*conclusion*
			case	dolphin	0	0	0	1	0	1	1	1	1	0	1	mammal	mammal. fish

order added	Go to if true	Go to if false	Rule no	*name*	*hair*	*feathers*	*eggs*	*milk*	*airborne*	*aquatic*	*backbone*	*breathes*	*fins*	*no of legs*	*tail*	*target*	*conclusion*
1	2	2	1					= 1									mammal
2	3	4	2							= 1							fish
5	4	4	3								= 0	= 0					
3	5	5	4			is 1											bird
4	6	6	5								= 0	= 0		= 0			mollusc
6	exit	exit	6								= 0	= 0					mollusc

FIGURE 6.11 The rule trace for a dolphin, giving both *mammal* and *fish*.

We need to add another stopping rule to stop the conclusion *fish*. On the *rule builder* worksheet when we first click *run case*, the conclusion *mammal* comes up (Figure 6.12), because the first rule in the knowledge base is the rule for a *mammal* and the cornerstone case for the *mammal* rule; the aardvark is also shown.

		name	hair	feathers	eggs	milk	airborne	aquatic	backbone	breathes	fins	no of legs	tail	target	conclusion	
run case	cornerstone	aardvark	1	0	0	1	0	0	1	1	0	4	0	mammal		final output: mammal
	current case	dolphin	0	0	0	1	0	1	1	1	1	0	1	mammal	mammal	*final output: mammal. fish*
		name	*hair*	*feathers*	*eggs*	*milk*	*airborne*	*aquatic*	*backbone*	*breathes*	*fins*	*no of legs*	*tail*	*target*	*conclusion*	
	new rule	*operator*	*operator*	*operator*	*operator*	*operator*	*operator*	*operator*	*operator*	*operator*	*operator*	*operator*	*operator*	*operator*	enter new conclusion here	
		value	*value*	*value*	*value*	*value*	*value*	*value*	*value*	*value*	*value*	*value*	*value*	*value*		

FIGURE 6.12 The conclusion *mammal* and the aardvark cornerstone case after *run* case is first clicked for the dolphin case.

When *run case* is clicked again, the *fish* rule fires as shown in Figure 6.13 and again the cornerstone case for this rule is shown.

		name	hair	feathers	eggs	milk	airborne	aquatic	backbone	breathes	fins	no of legs	tail	target	conclusion	
run case	cornerstone	bass	0	0	1	0	0	1	1	0	1	0	1	fish		final output: fish
	current case	dolphin	0	0	0	1	0	1	1	1	1	0	1	mammal	fish	*final output: mammal. fish*
		name	*hair*	*feathers*	*eggs*	*milk*	*airborne*	*aquatic*	*backbone*	*breathes*	*fins*	*no of legs*	*tail*	*target*	*conclusion*	
	new rule	*operator*	*operator*	*operator*	*operator*	*operator*	*operator*	*operator*	*operator*	*operator*	*operator*	*operator*	*operator*	*operator*	enter new conclusion here	
		value	*value*	*value*	*value*	*value*	*value*	*value*	*value*	*value*	*value*	*value*	*value*	*value*		

FIGURE 6.13 The *fish* conclusion and its cornerstone when run case is clicked a second time.

This is the rule that needs a stopping rule added to prevent the conclusion *fish* being given. The obvious stopping rule uses the condition *milk is 1.* In this demonstrator if the conclusion field is empty, then it is assumed that the rule is only a stopping rule, whereas if the rule has a conclusion, it is first added as a stopping rule and then it is considered as a candidate new rule giving the specified conclusion, as shown in the example above in Figure 6.8. Here no conclusion is included, so the rule shown below in Figure 6.14 is taken to be purely a stopping rule.

		name	hair	feathers	eggs	milk	airborne	aquatic	backbone	breathes	fins	no of legs	tail	target	conclusion	
run case	cornerstone	bass	0	0	1	0	0	1	1	0	1	0	1	fish		final output: fish
	currentcase	dolphin	0	0	0	1	0	1	1	1	1	0	1	mammal	fish	*final output: mammal. fish*
		name	*hair*	*feathers*	*eggs*	*milk*	*airborne*	*aquatic*	*backbone*	*breathes*	*fins*	*no of legs*	*tail*	*target*	*conclusion*	
	new rule	*operator*	*operator*	*operator*	*operator*	*is*	*operator*	*operator*	*operator*	*operator*	*operator*	*operator*	*operator*	*operator*		
		value	*value*	*value*	*value*	1	*value*	*value*	*value*	*value*	*value*	*value*	*value*	*value*		

FIGURE 6.14 A rule to stop *fish* being a conclusion for a dolphin.

Knowledge acquisition event: The next error is that a duck is classified as *fish* as well as a *bird*, as again the single condition *aquatic* = *1* is too broad a definition of a *fish*. For this case when we click *run case* the first rule that fires gives the incorrect conclusion *fish*, so we add another stopping rule for *fish*, which is automatically added as the last of its stopping rules. The rules and rule trace for a duck are shown in Figure 6.15.

			case	name	hair	feathers	eggs	milk	airborne	aquatic	backbone	breathes	fins	no of legs	tail	target	conclusion
			case	duck	0	1	1	0	1	1	1	1	0	2	1	bird	bird

order added	Go to if true	Go to if false	Rule no	name	hair	feathers	eggs	milk	airborne	aquatic	backbone	breathes	fins	no of legs	tail	target	conclusion
1	2	2	1					= 1									mammal
2	3	6	2							= 1							fish
5	6	4	3								= 0	= 0					
7	6	5	4					is 1									
8	6	6	5			is 1											
3	7	7	6			is 1											bird
4	8	8	7								= 0	= 0		= 0			mollusc
6	exit	exit	8								= 0	= 0					mollusc

FIGURE 6.15 Rules and the rule trace for a duck after a rule to stop a duck being classified as a *fish* is added.

Three stopping rules have been added to the overgeneralised *fish* rule: for the crab case, both a stopping rule and new rule to give the *mollusc* conclusion were added, while for the dolphin and duck, the right conclusions were already given but the wrong extra conclusion, *fish*, had to be stopped for these cases.

Knowledge acquisition event: The next error is for a flea. We start with the same rule for a flea as with SCRDR (Figure 5.14). The cornerstone cases are tested and none fire so the rule is added at the end of the list of rules.

Knowledge acquisition event: The next error is for a frog which as in SCRDR is misclassified as a *fish*, so we start with the same refinement rule added for SCRCR (Figure 5.15). As above, this is first added as stopping rule, and then the rule giving the conclusion is tested against the cornerstone cases before being added as a new rule. Again, the new rule will be added at the end, so all the cases for earlier rules will be tested against the new rule. The chicken cornerstone case satisfies the rule as *milk* = *0* and *breathes* = *1* apply to a chicken, so this cornerstone case is now shown as in Figure 6.16.

		name	hair	feathers	eggs	milk	airborne	aquatic	backbone	breathes	fins	no of legs	tail	target	conclusion	
run case	cornerstone	chicken	0	1	1	0	1	0	1	1	0	2	1	bird		final output: bird
	current case	frog	0	0	1	0	0	1	1	1	0	4	0	amphibian	amphibian	

	name	hair	feathers	eggs	milk	airborne	aquatic	backbone	breathes	fins	no of legs	tail	target	conclusion
new rule	operator	operator	operator	operator	=	operator	operator	operator	=	operator	operator	operator	operator	amphibian
	value	value	value	value	0	value	value	value	1	value	value	value	value	

FIGURE 6.16 A stopping rule that prevents a frog being classified as *fish* is inadequate as a rule for an *amphibian* as it applies to the cornerstone chicken.

To exclude the chicken we add the further condition *feathers* = *0* and the next cornerstone case that can fire the evolving rule is shown in Figure 6.17.

		name	hair	feathers	eggs	milk	airborne	aquatic	backbone	breathes	fins	no of legs	tail	target	conclusion	
run case	cornerstone	flea	0	0	1	0	0	0	0	1	0	6	0	insect		final output: insect
	current case	frog	0	0	1	0	0	1	1	1	0	4	0	amphibian	amphibian	

	name	hair	feathers	eggs	milk	airborne	aquatic	backbone	breathes	fins	no of legs	tail	target	conclusion
new rule	operator	operator	=	operator	=	operator	operator	operator	=	operator	operator	operator	operator	amphibian
	value	value	0	value	0	value	value	value	1	value	value	value	value	

FIGURE 6.17 The amphibian rule, after the chicken cornerstone is blocked, will still fire the flea cornerstone.

To exclude the flea cornerstone case, we add the further condition *backbone = 1*. The rule now excludes all the cornerstone cases and the rules are as shown in Figure 6.18. Note that the last rule (no 11) includes 2 extra conditions to exclude the flea and chicken cornerstones compared to rule 6 added as a stopping rule.

				name	hair	feathers	eggs	milk	airborne	aquatic	backbone	breathes	fins	no of legs	tail	target	conclusion
			case	frog	0	0	1	0	0	1	1	1	0	4	0	amphibian	amphibian
order added	**Go to if true**	**Go to if false**	**Rule no**	***name***	***hair***	***feathers***	***eggs***	***milk***	***airborne***	***aquatic***	***backbone***	***breathes***	***fins***	***no of legs***	***tail***	***target***	***conclusion***
1	2	2	1					= 1									mammal
2	3	7	2							= 1							fish
5	7	4	3								= 0	= 0					
7	7	5	4					is 1									
8	7	6	5			is 1											
10	7	7	6					= 0				= 1					
3	8	8	7			is 1											bird
4	9	9	8								= 0	= 0		= 0			mollusc
6	10	10	9								= 0	= 0					mollusc
9	11	11	10				= 1				= 0	= 1					insect
11	exit	exit	11			= 0		= 0			= 1	= 1					amphibian

FIGURE 6.18 The knowledge base corrected to classify a frog. A stopping rule is added so a frog is not classified as a fish and then a rule to classify it as an amphibian is added. This rule needed four conditions to exclude all cornerstones.

Knowledge acquisition event: The next case is for a pitviper. It is misclassified as an *amphibian*, but the same *reptile* rule as in Figure 5.17 is sufficient as both a stopping rule and a new rule, with no cornerstone cases firing.

Knowledge acquisition event: The next error is that a porpoise is classified as a *reptile* as well as a *mammal*, because of the rule we have just added (Figure 6.19).

		name	hair	feathers	eggs	milk	airborne	aquatic	backbone	breathes	fins	no of legs	tail	target	conclusion	
run case	cornerstone	pitviper	0	0	1	0	0	0	1	1	0	0	1	reptile		final output: reptile
	current case	porpoise	0	0	0	1	0	1	1	1	1	0	1	mammal	reptile	*final output: mammal. reptile*

	name	hair	feathers	eggs	milk	airborne	aquatic	backbone	breathes	fins	no of legs	tail	target	conclusion
new rule	operator	operator	operator	operator	operator	operator	operator	operator	operator	operator	operator	operator	operator	enter new conclusion here
	value	value	value	value	value	value	value	value	value	value	value	value	value	

FIGURE 6.19 The rule added for a pitviper misclassifies a porpoise.

The obvious stopping rule condition to add is *milk = 1*, and the resultant rules are shown in Figure 6.20. The rule trace shows that two identical stopping rules (but using a different operator: milk is 1 and milk = 1, for demonstration purposes) have to fire in order to stop a dolphin being given a *fish* or a *reptile* conclusion.

				name	*hair*	*feathers*	*eggs*	*milk*	*airborne*	*aquatic*	*backbone*	*breathes*	*fins*	*no of legs*	*tail*	*target*	*conclusion*
			case	porpoise	0	0	0	1	0	1	1	1	1	0	1	mammal	mammal
order added	Go to if true	Go to if false	Rule no	*name*	*hair*	*feathers*	*eggs*	*milk*	*airborne*	*aquatic*	*backbone*	*breathes*	*fins*	*no of legs*	*tail*	*target*	*conclusion*
1	2	2	1					= 1									mammal
2	3	7	2							= 1							fish
5	7	4	3								= 0	= 0					
7	7	5	4					is 1									
8	7	6	5			is 1											
10	7	7	6					= 0				= 1					
3	8	8	7			is 1											bird
4	9	9	8								= 0	= 0		= 0			mollusc
6	10	10	9								= 0	= 0					mollusc
9	11	11	10				= 1				= 0	= 1					insect
11	12	13	11			= 0		= 0			= 1	= 1					amphibian
12	13	13	12								= 1	= 1		= 0	= 1		
13	14	exit	13								= 1	= 1		= 0	= 1		reptile
14	exit	exit	14					= 1									

FIGURE 6.20 The knowledge base after a stopping rule was added so that a porpoise was not classified as a reptile.

Knowledge acquisition event: As with SCRDR, the next case needing a rule is a scorpion and we add the same rule as SCRDR (Figure 5.18). However, when cornerstone cases are tested the flea case fires this rule as shown in Figure 6.21.

		name	*hair*	*feathers*	*eggs*	*milk*	*airborne*	*aquatic*	*backbone*	*breathes*	*fins*	*no of legs*	*tail*	*target*	*conclusion*	
run case	cornerstone	flea	0	0	1	0	0	0	0	1	0	6	0	insect		final output: insect
	current case	scorpion	0	0	0	0	0	0	0	1	0	8	1	mollusc	mollusc	

	name	*hair*	*feathers*	*eggs*	*milk*	*airborne*	*aquatic*	*backbone*	*breathes*	*fins*	*no of legs*	*tail*	*target*	*conclusion*
new rule	*operator*	*operator*	*operator*	*operator*	*operator*	*operator*	*operator*	=	=	*operator*	>=	*operator*	*operator*	mollusc
	value	*value*	*value*	*value*	*value*	*value*	*value*	0	1	*value*	4	*value*	*value*	

FIGURE 6.21 The scorpion rule developed so far would misclassify a flea.

It is difficult to know what condition to add to the rule as we believe that *molluscs* probably lay eggs, and we doubt the difference between 6 and 8 legs is significant. As discussed above, we have already discovered that a scorpion is not a *mollusc* or an *insect*, but an *arachnid*, and that *arachnids* have three body sections, so in this knowledge base we use the condition *tail = 1* as perhaps representing a third body section as shown in Figure 6.22. This is successful in excluding all cornerstone cases.

		name	hair	feathers	eggs	milk	airborne	aquatic	backbone	breathes	fins	no of legs	tail	target	conclusion	
run case	cornerstone	flea	0	0	1	0	0	0	0	1	0	6	0	insect		final output: insect
	current case	scorpion	0	0	0	0	0	0	0	1	0	8	1	mollusc	mollusc	
new rule		name	hair	feathers	eggs	milk	airborne	aquatic	backbone	breathes	fins	no of legs	tail	target	conclusion	
		operator	operator	operator	operator	operator	operator	operator	=	=	operator	>=	=	operator	mollusc	
		value	value	value	value	value	value	value	0	1	value	4	1	value		

FIGURE 6.22 The *mollusc* rule for a scorpion after a tail condition is added, which excludes all cornerstones.

Knowledge acquisition event: The next error, as with SCRDR, is for a seasnake which is classified as a *fish*. We add the same rule as for SCRDR (Figure 5.19) with conditions *backbone* = *1* and *fins* = *0*. While perfectly fine for a refinement rule in SCRDR or a stopping rule, clearly many other animals have backbones but not fins. Automatic testing of the cornerstones proceeds, and the first cornerstone case which fires this rule is for a *mammal* shown in Figure 6.23.

		name	hair	feathers	eggs	milk	airborne	aquatic	backbone	breathes	fins	no of legs	tail	target	conclusion	
run case	cornerstone	aardvark	1	0	0	1	0	0	1	1	0	4	0	mammal		final output: mammal
	current case	seasna ke	0	0	0	0	0	1	1	0	0	0	1	reptile	reptile	
new rule		name	hair	feathers	eggs	milk	airborne	aquatic	backbone	breathes	fins	no of legs	tail	target	conclusion	
		operator	operator	operator	operator	operator	operator	operator	=	operator	=	operator	operator	operator	reptile	
		value	value	value	value	value	value	value	1	value	0	value	value	value		

FIGURE 6.23 The initial *reptile* rule added for a seasnake would misclassify a *mammal*.

The obvious choice to exclude a mammal is to add *milk* = *0* although this might not be the most general rule to exclude other non-reptiles. We add this condition and the next cornerstone case is for a chicken (Figure 6.24).

		name	hair	feathers	eggs	milk	airborne	aquatic	backbone	breathes	fins	no of legs	tail	target	conclusion	
run case	cornerstone	chicken	0	1	1	0	1	0	1	1	0	2	1	bird		final output: bird
	current case	seasna ke	0	0	0	0	0	1	1	0	0	0	1	reptile	reptile	
new rule		name	hair	feathers	eggs	milk	airborne	aquatic	backbone	breathes	fins	no of legs	tail	target	conclusion	
		operator	operator	operator	operator	=	operator	operator	=	operator	=	operator	operator	operator	reptile	
		value	value	value	value	0	value	value	1	value	0	value	value	value		

FIGURE 6.24 The *reptile* rule for a seasnake with a milk condition added.

We next add *no of legs* = *0*. No further cornerstone cases are shown to the user and we accept the rule. In fact, the pitviper also fires the new rule, but it is not shown because the conclusion is correct (Figure 6.25).

Knowledge acquisition event: As with SCRDR, the next error is that a slug is classified as an *insect* rather than a *mollusc*, so we add the same rule

		name	hair	feathers	eggs	milk	airborne	aquatic	backbone	breathes	fins	no of legs	tail	target	conclusion	
run case	cornerstone	chicken	0	1	1	0	1	0	1	1	0	2	1	bird		final output: bird
	current case	seasnake	0	0	0	0	0	1	1	0	0	0	1	reptile	reptile	
		name	hair	feathers	eggs	milk	airborne	aquatic	backbone	breathes	fins	no of legs	tail	target	conclusion	
	new rule	operator	operator	operator	operator	=	operator	operator	=	operator	=	=	operator	operator	reptile	
		value	value	value	value	0	value	value	1	value	0	0	value	value		

FIGURE 6.25 The final seasnake rule that excludes all cornerstones.

as for SCRDR using *no of legs* = *0* (Figure 5.20). This is fine as a stopping rule, but obviously too general for a new *mollusc* rule as there are other creatures without legs. The first cornerstone case that fires the new rule is for a *fish*, but we try to exclude other cases as well by using both *fins* = *0* and *backbone* = *0*. This excludes other cases except that we note the clam cornerstone case fires this rule, but is not shown to the user, as the conclusion is correct (Figure 6.26).

		name	hair	feathers	eggs	milk	airborne	aquatic	backbone	breathes	fins	no of legs	tail	target	conclusion	
run case	cornerstone	bass	0	0	1	0	0	1	1	0	1	0	1	fish		final output: fish
	current case	slug	0	0	1	0	0	0	0	1	0	0	0	mollusc	mollusc	
		name	hair	feathers	eggs	milk	airborne	aquatic	backbone	breathes	fins	no of legs	tail	target	conclusion	
	new rule	operator	operator	operator	operator	operator	operator	operator	operator	operator	operator	=	operator	operator	mollusc	
		value	value	value	value	value	value	value	value	value	value	0	value	value		

FIGURE 6.26 This overgeneralised *mollusc* rule would misclassify *fish*.

Knowledge acquisition event: As with SCRDR the next error is that a tortoise is classified as an *amphibian* rather than a *reptile*. We try the same rule as SCRDR (Figure 5.21), which seems fairly specific, but find that it also applies to birds when automatically tested against the cornerstones (Figure 6.27).

		name	hair	feathers	eggs	milk	airborne	aquatic	backbone	breathes	fins	no of legs	tail	target	conclusion	
run case	cornerstone	chicken	0	1	1	0	1	0	1	1	0	2	1	bird		final output: bird
	current case	tortoise	0	0	1	0	0	0	1	1	0	4	1	reptile	reptile	
		name	hair	feathers	eggs	milk	airborne	aquatic	backbone	breathes	fins	no of legs	tail	target	conclusion	
	new rule	operator	operator	operator	=	operator	operator	operator	=	=	operator	operator	=	operator	reptile	
		value	value	value	1	value	value	value	1	1	value	value	1	value		

FIGURE 6.27 A rule for a tortoise that would misclassify a chicken.

We add the condition *feathers* = *0* and the other cases are excluded (although the rule fired appropriately for the pitviper case). The final knowledge base is shown in Figure 6.28 and the rule trace for a pitviper is shown. It can be seen there are now three rules which fire all giving the same correct conclusion. No further cases are interpreted incorrectly, so the knowledge base is complete.

				name	hair	feathers	eggs	milk	airborne	aquatic	backbone	breathes	fins	no of legs	tail	target	conclusion
			c case	pitviper	0	0	1	0	0	0	1	1	0	0	1	reptile	reptile. reptile. reptile
order added	**Go to if true**	**Go to if false**	**Rule no**	***name***	***hair***	***feathers***	***eggs***	***milk***	***airborne***	***aquatic***	***backbone***	***breathes***	***fins***	***no of legs***	***tail***	***target***	***conclusion***
1	2	2	1					= 1									mammal
2	3	8	2							= 1							fish
5	8	4	3								= 0	= 0					
7	8	5	4					is 1									
8	8	6	5			is 1											
10	8	7	6					= 0				= 1					
16	8	8	7								= 1		= 0				
3	9	9	8			is 1											bird
4	10	10	9								= 0	= 0		= 0			mollusc
6	11	11	10								= 0	= 0					mollusc
9	12	13	11				= 1				= 0	= 1					insect
18	13	13	12											= 0			
11	14	16	13			= 0		= 0			= 1	= 1					amphibian
12	16	15	14								= 1	= 1		= 0	= 1		
20	16	16	15				= 1				= 1	= 1			= 1		
13	17	18	16								= 1	= 1		= 0	= 1		reptile
14	18	18	17					= 1									
15	19	19	18								= 0	= 1		>= 4	= 1		mollusc
17	20	20	19					= 0			= 1		= 0	= 0			reptile
19	21	21	20								= 0		= 0	= 0			mollusc
21	exit	exit	21			= 0	= 1				= 1	= 1			= 1		reptile

FIGURE 6.28 The complete MCRDR KB giving single conclusions for the Animal dataset.

6.3 DISCUSSION: MCRDR FOR SINGLE CLASSIFICATION

The MCRDR knowledge base in Figure 6.28 has 21 rules; 12 of these were rules giving conclusions, the other 9 were stopping rules. Of the stopping rules six were added automatically at the same time as a new rule was added; only three were added simply as stopping rules and two of these were added for a single case, the dolphin. Overall there were 14 knowledge acquisition events; that is, there were 14 cases for which corrections or additions had to be made – 12 where a conclusion was given and 2 for stopping rules only. In contrast 13 knowledge acquisition events, each adding a single rule, were needed for the SCRDR knowledge base in Chapter 5. For the 13 rules in the SCRDR knowledge base, 28 conditions were used, an average of 2.15 conditions per rule. For the 21 rules in the MCRDR knowledge base, 53 conditions were used an average of 2.52 per rule. If we look only at rules giving conclusions there are 2.99 conditions per MCRDR rules. It seems reasonable to make this comparison, since for each case for MCRDR we started with the same rule as SCRDR, only adding conditions as required to deal with cornerstone cases. Clearly, and as would be expected, MCRDR rules, all at the top level, tend to have more conditions to make them sufficiently precise, whereas fewer conditions are need for SCRDR rules because of the refinement structure.

Although rule addition for MCRDR seems more complex because a number of cornerstone cases may need to be excluded, more precise rules are being added. With SCRDR what might seem a general rule may be added as refinement rule rather than at the top level and may have to be written multiple times – although as we have discussed, in practice this does not seem to be a major problem. There is also some risk of this occurring with MCRDR, if refinement rules are used rather than stopping rules plus new rules, but at least all rules at the top level will be evaluated for a case, rather than just the first top level rule that fires as with SCRDR. The task of making sufficiently precise MCRDR rules is the same problem as with any knowledge base, where all rules are candidates to fire, but with MCRDR the problem is managed systematically with the cornerstone cases and RDR incremental development. Over time as cases are seen, rules will become sufficiently precise, and as noted in Appendix 1 over 800 MCRDR knowledge bases have been incrementally developed by Pacific Knowledge Systems customers. Experience also shows that even with potentially thousands of cornerstone cases, sufficiently precise rules are arrived at after the expert considers only a few cases, and as we have already noted the median time for the whole knowledge acquisition process for a case is only a minute or two (Compton et al. 2011, Compton 2013).

RDR knowledge bases are built by adding a rule when a new case being processed requires a rule, and only that case and past cornerstones are considered, resulting in a risk that cases that did not require a rule may be misclassified by later rules. If we look at the accuracy of the two knowledge bases developed so far on the training cases seen in building the rules, we find that the SCRDR knowledge bases process all the training cases correctly, but MCRDR makes an error for the newt case. This is because the newt had been correctly classified as an *amphibian* by the evolving rules and so did not require a rule and did not become a cornerstone case for a new rule. It was then misclassified when the final rule for a tortoise was added, but this was not flagged because it was not a cornerstone, just a case that had been classified correctly. This could also occur with SCRDR, but didn't in this study. One solution to this would be to keep all the cases processed, rather than just the cases which prompted the addition of a rule. This is attractive, but it is probably not practical to run inference on perhaps millions of cases – and certainly would not be practical with Excel_RDR!

As will be discussed in more detail, a critical aspect in deploying RDR is to keep monitoring cases and adding rules when necessary, until the user is satisfied the knowledge base is sufficiently accurate. This issue is not really discussed in the literature, but presumably the conventional approach is to build a knowledge base and then evaluate it on test cases and start using it routinely when it is deemed sufficiently accurate. The difference with RDR is that there is a way to do this incrementally as far as the user wishes to go, while the system is in actual use.

A final question is how the actual knowledge in the SCRDR knowledge base compares to that in a MCRDR knowledge base. To make this comparison an SCRDR knowledge base would have to be converted to flat rules. A rule with a number of refinement rules would be converted to multiple rules, each containing the conditions of the rule and its ancestors (if the rule is not the top rule) plus the negation of one condition from each of its refinement rules. This recursively applies to each refinement rule. The 13 rules shown in Figure 5.22 and Figure 5.23 become 19 rules with this conversion. If one similarly merges MCRDR stopping rules with their parent rule we end up with 18 rules. The numbers are similar because of the shallow depth of correction in SCRDR. As Colomb and Chung have noted, converting from a large SCRDR to a decision table (flat rules) can result in a huge blow-out in the number of rules and many of these rules are irrelevant, they do not apply to any cases and are just the result of the conversion (Colomb 1999). With MCRDR, flat (or rather composite) rules are built directly for real cases.

6.4 ACTUAL MULTIPLE CLASSIFICATION WITH MCRDR

In the Animal example so far, we have used MCRDR to provide single classification. The knowledge base will now be expanded to provide a further classification based on where the animal lives. For example, in the database there are mammals that live only on land, only in water, only in the air or in various combinations of these three environments. In a normal application we probably would have added these habitat rules at the same time as the species rules, as each case was considered.

Knowledge acquisition event: We start again with the first case. When adding a rule to give an extra conclusion we keep clicking *run case* going through the various conclusions that have been made until the conclusion field is empty as shown Figure 6.29.

		name	hair	feathers	eggs	milk	airborne	aquatic	backbone	breathes	fins	no of legs	tail	target	conclusion	
run case	cornerstone															
	current case	aardvark	1	0	0	1	0	0	1	1	0	4	0	mammal		final output: mammal
		name	hair	feathers	eggs	milk	airborne	aquatic	backbone	breathes	fins	no of legs	tail	target	conclusion	
	new rule	operator	operator	operator	operator	operator	operator	operator	operator	operator	operator	operator	operator	operator	enter new conclusion here	
		value	value	value	value	value	value	value	value	value	value	value	value	value		

FIGURE 6.29 An empty conclusion has to be shown, before an extra conclusion rule can be added.

So far, the *final output* for this case is mammal, but since the conclusion cell is now empty a rule with another conclusion can now be added. We decide that if an animal is not aquatic or airborne (*aquatic = 0* and *airborne = 0)* then it *lives only on land.* Previously when a cornerstone case fired a rule, we added extra conditions to the rule to ensure the new rule conclusion was not given for those cases. In adding an extra different type of conclusion we are now happy that this be added to some cornerstone cases where appropriate and we accept its addition to cornerstone cases: clam, flea, pitviper, scorpion, slug and tortoise. Accepting that a clam *lives only on land* highlights a further possible error in this dataset beyond the scorpion and seasnake as we had assumed clams were aquatic – but we used the data as downloaded from UCIrvine. We end up with the knowledge base in Figure 6.30 showing the rule trace for a tortoise.

Knowledge acquisition event: We next construct a rule for the bass, a *fish*, with the conclusion that it *lives only in water.* We use conditions

				name	hair	feathers	eggs	milk	airborne	aquatic	backbone	breathes	fins	no of legs	tail	target	conclusion
			case	tortoise	0	0	1	0	0	0	1	1	0	4	1	reptile	reptile. lives only on land
order added	Go to if true	Go to if false	Rule no	name	hair	feathers	eggs	milk	airborne	aquatic	backbone	breathes	fins	no of legs	tail	target	conclusion
1	2	2	1					= 1									mammal
2	3	8	2							= 1							fish
5	8	4	3								= 0	= 0					
7	8	5	4					is 1									
8	8	6	5			is 1											
10	8	7	6					= 0				= 1					
16	8	8	7								= 1		= 0				
3	9	9	8			is 1											bird
4	10	10	9								= 0	= 0		= 0			mollusc
6	11	11	10								= 0	= 0					mollusc
9	12	13	11				= 1				= 0	= 1					insect
18	13	13	12											= 0			
11	14	16	13			= 0		= 0			= 1	= 1					amphibian
12	16	15	14								= 1	= 1		= 0	= 1		
20	16	16	15				= 1				= 1	= 1			= 1		
13	17	18	16								= 1	= 1		= 0	= 1		reptile
14	18	18	17					= 1									
15	19	19	18								= 0	= 1		>= 4	= 1		mollusc
17	20	20	19					= 0			= 1		= 0	= 0			reptile
19	21	21	20								= 0		= 0	= 0			mollusc
21	22	22	21			= 0	= 1				= 1	= 1			= 1		reptile
22	23	23	22						= 0	= 0							lives only on land

FIGURE 6.30 The knowledge base and trace showing the multiple conclusions, reptile and lives on land for a tortoise.

aquatic = *1* and *fins* = *1* to make sure we don't cover amphibians, etc. The rule would have applied to seals and other amphibious mammals, but no such animals were included in the cornerstone cases, so we did not have to consider this conclusion for any such cornerstone cases (Figure 6.31).

		name	hair	feathers	eggs	milk	airborne	aquatic	backbone	breathes	fins	no of legs	tail	target	conclusion	
run case	cornerstone															
	current case	bass	0	0	1	0	0	1	1	0	1	0	1	fish		final output: fish
		name	hair	feathers	eggs	milk	airborne	aquatic	backbone	breathes	fins	no of legs	tail	target	conclusion	
	new rule	operator	operator	operator	operator	operator	operator	=	operator	operator	=	operator	operator	operator	lives only in water	
		value	value	value	value	value	value	1	value	value	1	value	value	value		

FIGURE 6.31 Adding a rule to give a second conclusion for animals that live only in water.

Knowledge acquisition event: We next add a rule for a chicken. In fact, we want to add two rules for a chicken, as we want to specify it flies, but it also lives on land – but not only on land. We could combine flying and living on land as a single conclusion, but the point of MCRDR is to use multiple conclusions and potentially fewer rules. We first add a rule concluding *it flies*, based on the simple obvious condition *airborne* = *1*. No cornerstone cases fire this rule. We then add a rule to conclude *it lives on land*. We click through *run case* to give both *bird* and *it flies* before getting to an empty conclusion field so the new rule for *it lives on land sometimes* can be added. We decide that anything that breathes and has legs must live on land – at least some of the time, so we add these two conditions. The first cornerstone case that fires the proposed rule is the aardvark, which *lives only on land* (we assume) and since we do not want both *lives only on land* and *lives on land sometimes* as conclusions for a case, we add *milk* = *0* to exclude all the mammals already covered by the *lives only on land* (Figure 6.32).

		name	hair	feathers	eggs	milk	airborne	aquatic	backbone	breathes	fins	no of legs	tail	target	conclusion	
run case	cornerstone	aardvark	1	0	0	1	0	0	1	1	0	4	0	mammal		final output: mammal, lives only on land
	current case	chicken	0	1	1	0	1	0	1	1	0	2	1	bird	it lives on land sometimes	
		name	hair	feathers	eggs	milk	airborne	aquatic	backbone	breathes	fins	no of legs	tail	target	conclusion	
	new rule	operator	operator	operator	operator	=	operator	operator	operator	=	operator	>	operator	operator	it lives on land sometimes	
		value	value	value	value	0	value	value	value	1	value	0	value	value		

FIGURE 6.32 A proposed rule for creatures that live sometimes on land excluding mammals.

But this is still inadequate as this rule applies to the flea, which we assume *lives only on land* (Figure 6.33).

		name	hair	feathers	eggs	milk	airborne	aquatic	backbone	breathes	fins	no of legs	tail	target	conclusion		
run case	cornerstone	flea	0	0	1	0	0	0	0	1	0	6	0	insect		final output: insect.	lives only on land
	current case	chicken	0	1	1	0	1	0	1	1	0	2	1	bird	it lives on land sometimes		
		nam e	hair	feathers	eggs	milk	airborne	aquatic	backbone	breathes	fins	no of legs	tail	target	conclusion		
new rule		operator	operator	operator	operator	=	operator	operator	operator	=	operator	>	operator	operator	it lives on land sometimes		
		value	value	value	value	0	value	value	value	1	value	0	value	value			

FIGURE 6.33 Narrowing the rule further to cover only animals that live on land sometimes.

To exclude *lives only on land* the choice for extra conditions that apply to a chicken is between multiple legs (>0) and backbone, so we add *backbone = 1*. We accept this applying to a frog, but it also applies to the tortoise cornerstone case, for which *lives only on land* is the correct conclusion (Figure 6.34).

		name	hair	feathers	eggs	milk	airborne	aquatic	backbone	breathes	fins	no of legs	tail	target	conclusion		
run case	cornerstone	tortoise	0	0	1	0	0	0	1	1	0	4	1	reptile		final output: reptile.	lives only on land
	current case	chicken	0	1	1	0	1	0	1	1	0	2	1	bird	it lives on land sometimes		
		nam e	hair	feathers	eggs	milk	airborne	aquatic	backbone	breathes	fins	no of legs	tail	target	conclusion		
new rule		operator	operator	operator	operator	=	operator	operator	=	=	operator	>	operator	operator	it lives on land sometimes		
		value	value	value	value	0	value	value	1	1	value	0	value	value			

FIGURE 6.34 The rule to cover animals that live on land sometimes needs to exclude tortoise.

We add the condition *feathers = 1*, and this excludes all cornerstone cases where the conclusion would be inappropriate, but it is not a very elegant rule with conditions *feathers = 1, milk = 0, backbone = 1, breathes = 1 and no of legs > 0.* Note that we are adding conditions that apply to a chicken, but not to the cornerstone cases we are trying to exclude. Although we accepted the earlier version of the rule without *feathers = 1* should apply to the frog cornerstone case, this case is now also excluded.

Knowledge acquisition event: The next case for consideration is a crab, as it is correctly classified as a *mollusc* but none of the rules so far conclude it *lives only in water* (we assume). The following rule is tested against cornerstones (Figure 6.35).

		name	hair	feathers	eggs	milk	airborne	aquatic	backbone	breathes	fins	no of legs	tail	target	conclusion	
run case	cornerstone															
	current case	crab	0	0	1	0	0	1	0	0	0	4	0	mollusc		final output: mollusc
		name	hair	feathers	eggs	milk	airborne	aquatic	backbone	breathes	fins	no of legs	tail	target	conclusion	
new rule		operator	operator	operator	operator	operator	operator	=	operator	=	operator	operator	operator	operator	lives only in water	
		value	value	value	value	value	value	1	value	0	value	value	value	value		

FIGURE 6.35 A proposed rule to classify an aquatic *mollusc* as living only in water.

The cornerstone cases that are covered are the bass, a *fish* and a seasnake, a *reptile*. The extra conclusion is accepted for the seasnake, but the bass is not shown as it already has this conclusion.

Knowledge acquisition event: The next case is the duck. As it is a *bird, it flies* and *lives sometimes on land* – but it should also have the conclusion, *lives in water.* We add a very specific rule with conditions *aquatic = 1* and *feathers = 1.* No cornerstone cases fire this rule, so the rule in Figure 6.36 is added to the knowledge base.

		name	hair	feathers	eggs	milk	airborne	aquatic	backbone	breathes	fins	no of legs	tail	target	conclusion	
run case	cornerstone															
	current case	duck	0	1	1	0	1	1	1	1	0	2	1	bird		final output: bird. it flies. it lives on land sometimes
		name	hair	feathers	eggs	milk	airborne	aquatic	backbone	breathes	fins	no of legs	tail	target	conclusion	
	new rule	operator	operator	=	operator	operator	operator	=	operator	operator	operator	operator	operator	operator	lives in water	
		value	value	1	value	value	value	1	value	value	value	value	value	value		

FIGURE 6.36 The duck already has two location conclusions, but needs a *lives in water* conclusion also.

Knowledge acquisition event: The next case is the frog, which is classified as an *amphibian,* but is not classified as living on water and on land. We add the following very specific rule to give the conclusion *lives on land sometimes* (Figure 6.37)

		name	hair	feathers	eggs	milk	airborne	aquatic	backbone	breathes	fins	no of legs	tail	target	conclusion	
run case	cornerstone															
	current case	frog	0	0	1	0	0	1	1	1	0	4	0	amphibian		final output: amphibian
		name	hair	feathers	eggs	milk	airborne	aquatic	backbone	breathes	fins	no of legs	tail	target	conclusion	
	new rule	operator	operator	operator	operator	operator	operator	=	operator	=	operator	>	operator	operator	lives on land sometimes	
		value	value	value	value	value	value	1	value	1	value	0	value	value		

FIGURE 6.37 A very specific rule to conclude a frog *lives on land sometimes.*

This also applies to the duck cornerstone case, which is correct. We next need a conclusion that the frog *lives in water.* We start with the rule shown in Figure 6.38 – again a fairly specific guess.

		name	hair	feathers	eggs	milk	airborne	aquatic	backbone	breathes	fins	no of legs	tail	target	conclusion	
run case	cornerstone															
	current case	frog	0	0	1	0	0	1	1	1	0	4	0	amphibian		final output: amphibian. lives on land sometimes
		name	hair	feathers	eggs	milk	airborne	aquatic	backbone	breathes	fins	no of legs	tail	target	conclusion	
	new rule	operator	operator	operator	operator	operator	operator	=	operator	=	=	operator	operator	operator	lives in water	
		value	value	value	value	value	value	1	value	1	0	value	value	value		

FIGURE 6.38 A very specific rule to conclude a frog *lives in water.*

The only cornerstone case for which the rule fires is again the duck – for which the conclusion is correct, so it is not shown.

Knowledge acquisition event: The next interesting case is the gnat. Although it flies, we believe that all flying things land sometimes, so we add a very generic rule that anything that is airborne *lives sometimes on land.* If we had thought of this earlier, we would have added the rule earlier.

This applies to cornerstones, chicken and duck. This rule is also a nice example that every generalisation is likely to have exceptions that are hard to think of in advance. Albatrosses do land, but if we had a case of a young albatross, perhaps we would have needed an exception because apparently young albatrosses do not touch land for six years or more.

Knowledge acquisition event: The next problem case is that a kiwi is classified as *lives only on land* and *lives on land sometimes* as shown in Figure 6.39.

21	22	22	21			= 0	= 1				= 1	= 1			= 1		reptile
22	23	23	22						= 0	= 0							lives only on land
23	24	24	23							= 1			= 1				lives only in water
24	25	25	24						= 1								it flies
25	26	26	25			= 1		= 0			= 1	= 1		> 0			it lives on land sometimes
26	27	27	26							= 1		= 0					lives only in water

FIGURE 6.39 The rules that fire for a kiwi.

Since the kiwi does not fly it should be only *lives only on land* and we add a stopping rule that uses *airborne = 0* to stop *it lives on land sometimes*

No more cases need rules, so we end up with the following knowledge base, showing the rule trace for a kiwi (Figure 6.40).

				name	*hair*	*feathers*	*eggs*	*milk*	*airborne*	*aquatic*	*backbone*	*breathes*	*fins*	*no of legs*	*tail*	*target*	*conclusion*
			case	kiwi	0	1	1	0	0	0	1	1	0	2	1	bird	bird. lives only on land
order added	Go to if true	Go to if false	Rule no	*name*	*hair*	*feathers*	*eggs*	*milk*	*airborne*	*aquatic*	*backbone*	*breathes*	*fins*	*no of legs*	*tail*	*target*	*conclusion*
1	2	2	1					= 1									mammal
2	3	8	2							= 1							fish
5	8	4	3								= 0	= 0					
7	8	5	4					is 1									
8	8	6	5			is 1											
10	8	7	6					= 0				= 1					
16	8	8	7								= 1		= 0				
3	9	9	8			is 1											bird
4	10	10	9								= 0	= 0		= 0			mollusc
6	11	11	10								= 0	= 0					mollusc
9	12	13	11				= 1				= 0	= 1					insect
18	13	13	12											= 0			
11	14	16	13			= 0		= 0			= 1	= 1					amphibian
12	16	15	14								= 1	= 1		= 0	= 1		
20	16	16	15				= 1				= 1	= 1			= 1		
13	17	18	16								= 1	= 1		= 0	= 1		reptile
14	18	18	17					= 1									
15	19	19	18								= 0	= 1		>= 4	= 1		mollusc
17	20	20	19					= 0			= 1		= 0	= 0			reptile
19	21	21	20								= 0		= 0	= 0			mollusc
21	22	22	21			= 0	= 1				= 1	= 1			= 1		reptile
22	23	23	22						= 0	= 0							lives only on land
23	24	24	23							= 1			= 1				lives only in water
24	25	25	24						= 1								it flies
25	26	27	25			= 1		= 0			= 1	= 1		> 0			it lives on land sometimes
30	27	27	26						= 0								
26	28	28	27							= 1		= 0					lives only in water
27	29	29	28			= 1				= 1							lives in water
28	30	30	29							= 1		= 1		> 0			lives on land sometimes
29	exit	exit	30							= 1		= 1	= 0				lives in water

FIGURE 6.40 The complete knowledge base, giving both animal type and location conclusions.

We have added 10 more rules to correctly apply the 5 classifications: *lives only on land, it lives on land sometimes, lives only in water, lives in water* and *it flies.*

6.5 DISCUSSION

The examples above demonstrate how MCRDR can assign multiple conclusions. This is a simpler example than real world applications such as multiple interpretations in chemical pathology. In the example above where an animal lives is quite a different type of classification from the type of animal. However, the examples also suggest some of the issues that might arise if conclusions are simply added regardless of any relationship between them. The discussion of these issues here is based on past discussion with Lindsay Peters of Pacific Knowledge Systems and his direct experience of these issues with in-use MCRDR systems.

- The conclusions given for a swan are as follows: *bird, it flies, it lives on land sometimes, lives in water, lives on land sometimes, lives in water, it lives on land sometimes.* The classifications are all correct, but for most applications, it is not appropriate to simply pump out such output; not only should each concept appear only once, the output needs to be formatted, e.g. *a swan is a bird which lives in water, but lives on land sometimes.* MCRDR itself doesn't provide any capability to structure the output; as far as MCRDR is concerned the output is simply a number of independent concepts. Appropriate combining and formatting of multiple conclusions is something that needs to be done with separate rules or outside MCRDR. This is discussed further in Chapter 7.
- If the same conclusion is made repeatedly, but the user wants to change or stop that conclusion, how is this best done? The same stopping or refinement rule can be automatically added under each rule giving the wrong conclusion, but this involves testing the cornerstone case for each rule giving the wrong conclusion, to give a sufficiently precise rule to stop all these cases firing on the new stopping or refinement rule. This could have been done in Excel_MCRDR, but since the aim of the worked examples was to show exactly how RDR worked, it seemed better to make readers directly aware of the issue rather than mask it.

- The converse problem can also arise that although only one rule fires for a case, the user might want to stop or change the conclusion, not only for that case and rule, but for every rule that gives that conclusion. The user may wish to go further and stop a range of related conclusions, even though only one of the conclusions has been given by one rule for the case at hand. If this is needed, some sort of ad hoc extra functionality has to be included. We also note that none of these problems is unique to RDR, rather they are issues that have to be addressed with any rule-based system, but are less likely with RDR. In Chapter 7 we suggest a more general solution to such problems.
- In the example in Figure 6.10, rule 6 is more general than rule 5. This suggests that one could simply delete the more specific rule, but this is not always the case. For example, one might have the rules:

  ```
  IF A & B & C THEN X

  IF B & C & D THEN X
  ```

  The rules do not subsume each other but both would fire on a case with features A,B,C & D. Nor can one assume that the rules could be replaced with a rule just with the shared conditions, such as:

  ```
  IF B & C THEN X
  ```

- The examples presented here all use very simple concepts for the conclusion, so automatically adding the same stopping or refinement rule under each rule giving a particular conclusion would be fairly straightforward. However, in a real application the concepts may be text strings freely typed by the domain expert so the same concept/conclusion is expressed in slightly different ways in different rule conclusions. The same stopping or refinement rule cannot be added to such rules automatically, because it would become a significant text understanding problem to decide the conclusions were the same. On the other hand, obviously it is likely to irritate the expert to have to add the same stopping/refinement rule multiple times. The obvious solution is to give the domain expert the extra task of first checking and, if possible, selecting from a list of conclusions already used.

This itself may be irritating and ultimately there is no way of ensuring the expert will not just add another version of the same concept. The GRDR approach in Chapter 7 attempts to encourage the user to first use simple concept conclusions, with the longer free text expression of concepts and combination of concepts handled separately.

- Finally, and related to the last point, how does the user decide what concepts to use? In the examples we have given we used concepts such as: *lives only on land* and *lives on land sometimes* and also *lives in water* and *lives only in water.* With hindsight this was not necessary, because the qualifiers are implicit when multiple conclusions are given together. For example, if both conclusions *lives on land* and *lives in water* are given, it is obvious that the animal *lives on land sometimes* and conversely it is not true that it *lives only in water.* The rules evolved as they did because the first land rule concluded *lives only on land,* so it was then necessary to add a rule concluding *lives on land sometimes* and similarly the first water rule concluded *lives only in water.* Even though what we did was probably unnecessarily clumsy, it did not make the knowledge acquisition task more difficult, but perhaps slightly longer with a few extra rules. RDR adopts the approach that it makes knowledge building faster and easier to just go ahead without worrying too much about ontological considerations. Being more careful about the ontology will take extra effort, while the underlying philosophical problems still remain as was discussed in Chapter 2. Of course, we are not suggesting domain experts should be totally cavalier in their naming of concepts, but we are suggesting it is more productive to enable a domain expert to easily fix errors rather than hoping they can be prevented from making any.

6.6 SUMMARY

If a user is confident that only a single classification will be required, then SCRDR is appropriate. If on the other hand it is not clear that only a single classification is required, then it is better to use MCRDR. In developing the first industrial RDR system using SCRDR (Edwards et al. 1993), an important consideration was selecting subdomains that were unlikely to occur together in a typical pathology report, as otherwise many more rules might have been required with a single conclusion for each possible combination of conclusions. Another approach is to use multiple SCRDR knowledge bases as used by Dani et al. (2010), but this requires a clear demarcation between different subdomains, which is not always obvious in a domain like chemical pathology. As we have noted, the final formatting of MCRDR output with a number of conclusions needs to be managed outside the knowledge base, or as a special part of the knowledge base. The same applies when using multiple SCRDR knowledge bases. We note that we are identifying issues that apply to the practical industrial application of any knowledge-based system not just RDR, but are not normally discussed in the knowledge-based system literature.

In the example here we did not allow refinement rules, with any rule having perhaps many refinements forming an n-ary tree. Instead, whenever a rule gave a wrong conclusion, we added a stopping rule and a new rule to give the right conclusion. Although we have perhaps implied that it is better to have all rules at the top level like this, there are other considerations. If an application requires "why not" explanations, there might be an advantage in refinement rules, as one can then see why the child conclusion rather than parent conclusion was given. "Why not" explanations are little more obscure when a rule has many stopping rules attached to it.

In the Zoo example domain, there is no need for any initial data abstraction, but in a domain like chemical pathology, obviously numerical data needs to be abstracted to high, normal, low, undetectable, etc. with respect to the normal range for the analyte with perhaps further abstractions for borderline results. A result can be high or only borderline-high, or may be normal but borderline-high and the interpretation of these borderline results will depend more on other data than more clearly high or normal results. Further abstractions can be needed if temporal data is used – e.g. in chemical pathology there will be previous results as well as current results, so abstractions such as maximum, minimum, increasing, decreasing etc. are useful. These sorts of abstractions can be developed separately from the rules and technology like that of PKS provides a separate language for users to write functions to provide this sort of data abstraction. Clancey's discussion of Heuristic Classification while covering this type of function-based abstraction is more concerned with rules that make intermediate conclusions which are then used as conditions in further rules (Clancey 1985). This type of rule abstraction is not provided for directly with MCRDR or SCRDR but is sometimes implemented with some sort of preprocessor rule base in industrial RDR systems. It is discussed as a central feature of the GRDR approach in Chapter 7.

CHAPTER 7

General Ripple-Down Rules (GRDR)

7.1 BACKGROUND

Industrial Ripple-Down Rules (RDR) systems to date have used either Single Classification Ripple-Down Rules (SCRDR), Multiple Classification Ripple-Down Rules (MCRDR) or some bespoke variant of these. SCRDR and MCRDR are essentially used for classification, as is machine learning. This chapter is included for those interested in considering what is involved in developing a General Ripple-Down Rules (GRDR) system able to carry out a wider range of tasks than just classification, similar to other general-purpose rule systems – although they too are mainly used for classification. GRDR has not been used in industry although some of its features appear in various RDR systems in use in industry.

An essential feature of any general-purpose rule engine is repeat inference with facts being asserted and retracted and later inference depending on changes in the facts available until no further changes are made. There has been a range of work on repeat inference cycles with RDR, with conclusions depending on each other. Beydoun and Hoffmann developed Nested Ripple-Down Rules (NRDR) with a separate SCRDR knowledge base for each concept, where the rules in any of the knowledge bases can use conclusions from the other knowledge bases as rule conditions (Beydoun and Hoffman 1997, 1998, 1999, 2000, 2001). This is similar to a standard rules system in that it needs an explicit mechanism to prevent cycles occurring when a conclusion is asserted and then retracted by other rules and then reasserted by the original rules. Although the rules in each knowledge base are developed incrementally, because the knowledge bases are separate, the order of rule addition across the knowledge bases is not maintained, only within each single knowledge base.

A similar approach to NRDR was used in a commercial project cleansing Indian address data, where a different SCRDR tree was used for each field that had to be corrected (Dani et al. 2010). After the initial fields are specified, the expert then specifies the order in which the SCRDR trees for the various fields are processed and extra fields can be added at any stage. There do not appear to be any inference cycles in this method, but it is worth noting again that this was an industrial application and the developers won an award given when a piece of research brings in at least $10M of new business (See Appendix 1)

Multiple Classification Ripple-Round Rules (Bindoff and Kang 2011) not only used an MCRDR n-ary tree knowledge base but also maintained a graph structure that kept track of which rules need conclusions from other rules to fire and the status of these conclusions. Similar to NRDR it required a mechanism to prevent cycles. The method was validated using simulated data from an artificial configuration task.

Vazey and Richards describe a hybrid classification and configuration RDR system developed for a high-volume ICT call centre (Vazey 2007, Vazey and Richards 2006a). Advice generally cannot be provided on an initial fault report, but through an interactive process a full case description can be built up. This is a sort of configuration task, as the various aspects of the case have to make sense together, but the advice about different aspects of the case is provided through classification.

Pacific Knowledge Systems' RippleDown system allows repeat inference with an MCRDR structure, but only a limited number of cycles are allowed, essentially providing rules for data abstractions before rules for final conclusions[1].

The GRDR system approach described below is based on the proposal in Compton, Kim, and Kang (2014) following an earlier proposal (Compton, Cao, and Kerr 2004), and emerged from application studies on configuring ion chromatography and resource allocation (Ramadan et al. 1998b, Compton et al. 1998, Richards and Compton 1999, Compton and Richards 2000). The version of GRDR proposed in Compton, Kim, and Kang (2014) was based on MCRDR in that conclusions can be considered as independent Booleans. It was also similar to the MCRDR implementation in Chapter 6 in that a change in a conclusion for a case resulted in both a stopping rule and a new rule. In the version of GRDR we describe below, conclusion variables can only take one of a number of mutually exclusive values and we allow only SCRDR-style refinement rules. The essential features of the two approaches are the same, but the earlier proposal was based on MCRDR, while the proposal here is based on SCRDR. We do not claim that the GRDR system outlined below will work better than NRDR or these other more general RDR systems; there has not been enough evaluation or comparison of any of these various methods to make such claims. However, it should be noted that there are many Pacific Knowledge System knowledge bases in use with its version of MCRDR with a limited number of inference cycles.

7.2 KEY FEATURES OF GRDR

The key aspects of GRDR were outlined in Compton, Kim, and Kang (2014) and foreshadowed in Ramadan et al. (1998b), Richards and Compton (1999) and Compton and Richards (2000).

7.2.1 Linked Production Rules

As with SCRDR and MCRDR every rule has two "inference actions" as well as a "case action". A "case action" specifies what conclusion is asserted if the rule fires whereas the inference actions specify what rules are to be evaluated next depending on whether the rule is true or not.

7.2.2 No Retraction

To solve more complex problems, repeat inference (repeat traversals of the knowledge base) is essential, and in general, knowledge-based system shells provide for rules being traversed and re-traversed until facts are no longer changing. Secondly, during inference in these shells, facts can be both asserted and retracted. Any discussion of knowledge-based system (KBs) technology talks of the problems of cycles: a fact gets asserted which may lead to other rules firing which end up retracting the fact, so it is reasserted and so on. In contrast:

A central idea in GRDR is that during repeat inference cycles "data" or "facts" cannot be changed or retracted by a later inference step.

Of course, one can build a KBS to advise that data seems incorrect or problematic in some way. But this assessment of data is quite different from inference actually changing the data provided and then drawing conclusions from that changed data. It is a central assumption of the GRDR approach that it should be the responsibility of the data source to either change the data or to verify that the KBS's proposed change is correct.

If we take the same idea further, it should not be the responsibility of a rule to change a conclusion (retract the fact) asserted by another rule earlier in inference that it is unrelated to. Rather, if a rule is giving the wrong conclusion in some circumstance, then a refinement rule should be added so that it gives the correct conclusion; that is, retracting facts should not be the responsibility of inference, rather knowledge should be added to the system to ensure facts are not asserted inappropriately.

If a rule asserts a fact that is incorrect, it should be prevented from asserting the incorrect fact in the first place by adding a refinement rule or a stopping rule.

7.2.3 Inference Restarts After a Fact Is Asserted

With conventional rules systems, every time a rule fires and a fact is asserted or retracted all rules become candidates to fire and a conflict resolution strategy is used to determine which rule should be evaluated next. As discussed previously, conflict resolution strategies may include selecting a rule which has the most conditions, or which covers the most recently added fact, etc. In contrast:

GRDR inference simply returns to the first rule whenever a conclusion is made (i.e. a fact is asserted).

If a user decides an extra conclusion should be added to a case, they make this decision having considered all the conclusions added already and they are free to use one or more of these conclusions as a condition in the new rule. If the same case is run again later after the rule has been added, we expect the rule to fire and give the desired conclusion. But for this to happen then those same conditions that were used to create the rule must be available when inference reaches the newly added rule. That means inference must have repeated on the earlier rules, until no more conclusions are added – in exactly the same way as before the new rule was added. This is achieved simply by inference returning to the first rule each time a conclusion is asserted; that is, a rule is never evaluated until all possible conclusions from previous rules have been asserted. In a conventional system, after a fact is asserted, all rules become candidates to fire. For the same to happen with linked rules, inference needs to return to the first rule.

It may be possible to develop mechanisms similar to a RETE network (Forgy 1982) so that only earlier rules that might use a newly asserted conclusion are evaluated, but because of the linked production rule approach specifying the next rule to be evaluated, there is no requirement for general searching for a candidate rule.

Like any standard knowledge-based system, inference proceeds until no new conclusions are asserted.

Errors in conclusions are corrected in the order in which they occur.

When a case is run with GRDR a number of conclusions may be made. As noted, these conclusions are made one at a time and after each conclusion is asserted inference recommences at the first rule. This means that later conclusions can only depend on conclusions made earlier or the original data. Because of this dependence any errors made on a particular case should be corrected one at a time starting with the first, and then the case rerun and the next first error corrected. Correcting an earlier error may result in a later conclusion being changed, so this iterative approach, correcting the first error, going back to the start and correcting whatever is the new first error is essential.

7.2.4 GRDR Knowledge Base Structure

The version of GRDR we demonstrate here produces what is essentially a series of interspersed partial SCRDR knowledge bases for each concept.

The left-hand side of Figure 7.1 is the same as Figure 5.23, an SCRDR knowledge base; however, we have (arbitrarily) marked three cuts in the knowledge base. The right-hand side of Figure 7.1 shows the same rules but with some habitat rules where the cuts are. These rules are all part of the same knowledge base, with the rules added whenever they are required.

With the knowledge base on the left inference would proceed down the rules and inference would stop after a rule fired and none of its refinement rules fired; no further rules are evaluated. In the knowledge base on the right the rules are evaluated in order and as soon as a rule fires where none of its refinement rules fire inference returns to the first rule; that is, if none of the first block of *species* rules fires, but then one of the first block of *habitat* rules fires, inference returns to the first rule. The red *habitat* rules will then no longer fire as there is a *habitat* conclusion, so inference for the *species* rules will then be exactly the same as for the *species-only* rules on the left of Figure 7.1. However, if the conclusion from the *habitat* rule was used in a rule condition by one of the rules in the first block of *species* rules, then of course, the *species* inference conclusion might be different – but this is exactly the same as if the *habitat* conclusion had been part of the input data. This structure will become clearer with the example knowledge base below.

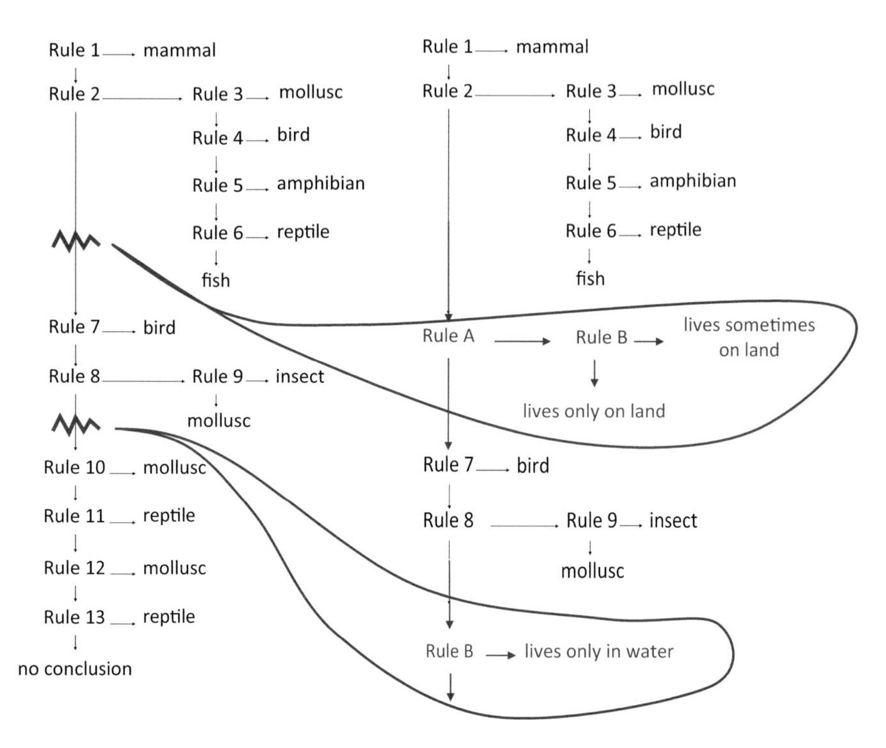

FIGURE 7.1 The knowledge base from Figure 5.23 separated into three sequential KBs.

7.2.5 The Final Conclusion

A knowledge-based system not only takes in data from outside itself, but it also outputs its conclusions back to the outside world. This may be an alarm, a report or perhaps a control signal to a machine. A GRDR system as outlined so far just outputs a number of separate conclusions, in the order in which they have been given, rather than these being combined in some way. They can be combined outside the knowledge base, but it is also possible to do this within the knowledge base. Excel_GRDR does this by providing for a final combination conclusion, as a simple example of what might be done, but the actual real-world application will determine what is required for the final output. In Excel_GRDR, as in SCRDR and MCRDR, there is column labelled "conclusion" which is the final conclusion to be output.

With GRDR when a conclusion is asserted for this final conclusion, there is no repeat inference, and this conclusion cannot be used as a condition in a rule.

The final conclusion is the conclusion to be output so when one of these output rules fires without a refinement, the inference action must be to exit the knowledge base, passing the output conclusion to the outside.

In this implementation the system simply puts final conclusion rules after the other rules. This means if we add a rule that gives an intermediate conclusion (and for which there will be repeat inference), the system pushes the final conclusion rules down and inserts the new rule before them. Perhaps it might have been more elegant to have a separate knowledge base, but this is an implementation issue. We should also note that having a final conclusion that is output and cannot be used for inference is not essential to GRDR in the same way as not allowing retraction etc. Inference could simply stop when no new facts were asserted.

7.3 EXCEL_GRDR DEMO

We use the same Animal example, but add two extra fields, *species* and *habitat* as shown below (Figure 7.2).

This sheet is for Cases you will evaluate with the RDR KB
Enter attribute names in the next row in place of attribute 1, 2 etc, and then Cases below

name	*hair*	*feathers*	*eggs*	*milk*	*airborne*	*aquatic*	*backbone*	*breathes*	*fins*	*no of legs*	*tail*	*species*	*habitat*	*target*	*conclusion*
aardvark	1	0	0	1	0	0	1	1	0	4	0			mammal	
antelope	1	0	0	1	0	0	1	1	0	4	1			mammal	
bass	0	0	1	0	0	1	1	0	1	0	1			fish	
bear	1	0	0	1	0	0	1	1	0	4	0			mammal	
boar	1	0	0	1	0	0	1	1	0	4	1			mammal	
buffalo	1	0	0	1	0	0	1	1	0	4	1			mammal	
calf	1	0	0		0	0	1	1	0	4	1			mammal	
carp	0	0	1	0	0	1	1	0	1	0	1			fish	
catfish	0	0	1	0	0	1	1	0	1	0	1			fish	

FIGURE 7.2 The Zoo data with two extra fields: species and habitat.

To demonstrate GRDR we also delete some data for some cases. We delete the *milk* value for a calf, hamster and hare and the *feathers* value for dove and duck.

Knowledge acquisition event: We click on the aardvark and choose to add a rule. As before this takes us to the rule builder sheet, and we click *run case*. As no conclusion is given, we see the following (Figure 7.3).

		name	*hair*	*feathers*	*eggs*	*milk*	*airborne*	*aquatic*	*backbone*	*breathes*	*fins*	*no of legs*	*tail*	*species*	*habitat*	*target*	*conclusion*
	cornerstone																
run case	current case	aardvark	1	0	0	1	0	0	1	1	0	4	0			mammal	
		name	*hair*	*feathers*	*eggs*	*milk*	*airborne*	*aquatic*	*backbone*	*breathes*	*fins*	*no of legs*	*tail*	*species*	*habitat*	*target*	*conclusion*
	new rule	*operator*	*operator*	*operator*	*operator*	*operator*	*operator*	*operator*	*operator*	*operator*	*operator*	*operator*	*operator*	*operator*	*operator*	*operator*	enter new conclusion here
		value	*value*	*value*	*value*	*value*	*value*	*value*	*value*	*value*	*value*	*value*	*value*	*value*	*value*	*value*	

FIGURE 7.3 The initial rule builder worksheet after clicking run case.

We want to add a conclusion for *species*, so we replace *conclusion* with *species* in the field above "enter new conclusion". Alternatively, we double click on the *species* field on the first grey row. Double-clicking on an attribute name in the first grey row, or entering an attribute name in the conclusion field at the end of the attribute names only works if there isn't a value for that case – as you can't add a new rule to give a conclusion where there is already one. But here, doing either to select *species* will produce the following (Figure 7.4).

	name	*hair*	*feathers*	*eggs*	*milk*	*airborne*	*aquatic*	*backbone*	*breathes*	*fins*	*no of legs*	*tail*	*species*	*habitat*	*target*	*conclusion*
cornerstone																
current case	aardvark	1	0	0	1	0	0	1	1	0	4	0			mammal	
	name	*hair*	*feathers*	*eggs*	*milk*	*airborne*	*aquatic*	*backbone*	*breathes*	*fins*	*no of legs*	*tail*		*habitat*	*target*	*species*
new rule	*operator*	*operator*	*operator*	*operator*	*operator*	*operator*	*operator*	*operator*	*operator*	*operator*	*operator*	*operator*		*operator*	*operator*	enter new species conclusion here
	value	*value*	*value*	*value*	*value*	*value*	*value*	*value*	*value*	*value*	*value*	*value*		*value*	*value*	

FIGURE 7.4 The rule builder sheet after a conclusion variable for the rule is selected.

The fields where we enter rule conditions are blanked out for *species* as *species* will be the conclusion of the rule we are going to construct. We construct the same rule as before, except since we are making a rule to give a conclusion about *species,* the conclusion name is now *species* and the rule from Figure 7.5 is:

```
IF milk =1 THEN species = mammal
```

	name	*hair*	*feathers*	*eggs*	*milk*	*airborne*	*aquatic*	*backbone*	*breathes*	*fins*	*no of legs*	*tail*	*species*	*habitat*	*target*	*conclusion*
cornerstone																
current case	aardvark	1	0	0	1	0	0	1	1	0	4	0			mammal	
	name	*hair*	*feathers*	*eggs*	*milk*	*airborne*	*aquatic*	*backbone*	*breathes*	*fins*	*no of legs*	*tail*		*habitat*	*target*	*species*
new rule	*operator*	*operator*	*operator*	*operator*	=	*operator*	*operator*	*operator*	*operator*	*operator*	*operator*	*operator*		*operator*	*operator*	mammal
	value	*value*	*value*	*value*	1	*value*	*value*	*value*	*value*	*value*	*value*	*value*		*value*	*value*	

IF milk = 1 THEN species = mammal

FIGURE 7.5 A rule for species = mammal.

Adding this rule takes us to the rules screen, where the rule is shown with the rule conclusion outlined in black on the rule row. To run the case the user clicks the *run* button and the conclusion is outlined in red (Figure 7.6). Note that the *go to if true* cell shows *restart* since whenever a rule fires which does not have a refinement, inference returns to the first rule. Since this rule asserts *species = mammal* which had already been asserted the first time this rule fired, the rule cannot fire again, and inference will go to the *go to if false* rule if *run* is clicked again.

This sheet contains the rules you have built; you can edit the case, but not the rules

				name	hair	feathers	eggs	milk	airborne	aquatic	backbone	breathes	fins	no of legs	tail	species	habitat	target	conclusion
run			case		1	0	0	1	0	0	1	1	0	4	0	mammal		mammal	
order added	Go to if true	Go to if false	Rule no	name	hair	feathers	eggs	milk	airborne	aquatic	backbone	breathes	fins	no of legs	tail	species	habitat	target	conclusion
1	restart	exit	1					=1								mammal			

FIGURE 7.6 The rules screen after adding the first rule, with the rule conclusion highlighted with a border in the rules and the case.

If we return to the *cases* screen and run the aardvark case we see the screen shown in Figure 7.7. The new conclusion added is outlined in red. This outlining is used, because in other cases the value for this attribute may have been part of the input data, and it may be helpful to the user to see what has been added by inference, rather than being part of the original data.

name	hair	feathers	eggs	milk	airborne	aquatic	backbone	breathes	fins	no of legs	tail	species	habitat	target	conclusion
aardvark	1	0	0	1	0	0	1	1	0	4	0	mammal		mammal	
antelope	1	0	0	1	0	0	1	1	0	4	1			mammal	
bass	0	0	1	0	0	1	1	0	1	0	1			fish	
bear	1	0	0	1	0	0	1	1	0	4	0			mammal	
boar	1	0	0	1	0	0	1	1	0	4	1			mammal	

FIGURE 7.7 The result of running the first rule on the aardvark case shown on the cases worksheet.

We then decide to add another rule for the aardvark case and click *add rule for this case*. It is important that when we want to add a rule for a case, we start with the original case and use *run case* to keep adding conclusions one by one until we get to the one we want to change (refine) or until we get to the red border around the empty conclusion field and can enter a further conclusion. Once a conclusion is added, at the next inference step it will be perceived as if it has come from outside the system, and so cannot be changed. Only a conclusion that is being asserted at a particular inference step can be changed by adding a refinement rule at that inference step; that is, rather than retracting it, the refinement rule prevents the wrong assertion being made. This means we have to step through each conclusion clicking *run case* until we get to the conclusion we want to change, or until no conclusion is given. Changing an earlier conclusion may also result in a later conclusion no longer being correct, so it would also require correction.

If we click *run case* on the rule builder screen, we get Figure 7.8.

	name	hair	feathers	eggs	milk	airborne	aquatic	backbone	breathes	fins	no of legs	tail	species	habitat	target	conclusion
cornerstone	aardvark	1	0	0	1	0	0	1	1	0	4	0	mammal		mammal	
current case	aardvark	1	0	0	1	0	0	1	1	0	4	0	mammal		mammal	
	name	hair	feathers	eggs	milk	airborne	aquatic	backbone	breathes	fins	no of legs	tail		habitat	target	species
new rule	operator	operator	operator	operator	operator	operator	operator	operator	operator	operator	operator	operator		operator	operator	enter new species conclusion here
	value	value	value	value	value	value	value	value	value	value	value	value		value	value	

FIGURE 7.8 Clicking run case on the rule builder screen, after one rule has been added.

Figure 7.3 showed an empty conclusion field when we clicked *run case* and we could then add a new rule. In contrast, after that first rule is added and we click *run case* on the rule builder screen we get Figure 7.8 where the *species* conclusion is highlighted. If appropriate we can now add a refinement rule to change this *species* conclusion. In this case we don't want to change this conclusion so we click *run case* again and get Figure 7.9 where *species* = *mammal* now becomes a fact that has been asserted and cannot be changed and the empty conclusion field is outlined showing that we can now add a new rule.

		name	hair	feathers	eggs	milk	airborne	aquatic	backbone	breathes	fins	no of legs	tail	species	habitat	target	conclusion
run case	cornerstone																
	current case	aardvark	1	0	0	1	0	0	1	1	0	4	0			mammal	
		name	hair	feathers	eggs	milk	airborne	aquatic	backbone	breathes	fins	no of legs	tail	species	habitat	target	conclusion
	new rule	operator	operator	operator	operator	operator	operator	operator	operator	operator	operator	operator	operator	operator	operator	operator	enter new conclusion here
		value	value	value	value	value	value	value	value	value	value	value	value	value	value	value	

FIGURE 7.9 Clicking run case a second time after Figure 7.8.

We double-click the grey field for *habitat* or enter *habitat* in the conclusion cell in the new rule section. We then get Figure 7.10.

	name	hair	feathers	eggs	milk	airborne	aquatic	backbone	breathes	fins	no of legs	tail	species	habitat	target	conclusion
cornerstone																
current case	aardvark	1	0	0	1	0	0	1	1	0	4	0	mamma		mammal	
	name	hair	feathers	eggs	milk	airborne	aquatic	backbone	breathes	fins	no of legs	tail	species		target	habitat
new rule	operator	operator	operator	operator	operator	operator	operator	operator	operator	operator	operator	operator	operator		operator	enter new habitat conclusion here
	value	value	value	value	value	value	value	value	value	value	value	value	value		value	

FIGURE 7.10 *Habitat* has been selected as the conclusion variable.

We enter the following rule for *habitat*, the same rule we used for MCRDR (Figure 7.11).

After adding this rule, we are taken to the rules screen which shows Figure 7.12.

	name	hair	feathers	eggs	milk	airborne	aquatic	backbone	breathes	fins	no of legs	tail	species	habitat	target	conclusion
cornerstone																
current case	aardvark	1	0	0	1	0	0	1	1	0	4	0	mammal		mammal	

	name	hair	feathers	eggs	milk	airborne	aquatic	backbone	breathes	fins	no of legs	tail	species	habitat	target	conclusion
new rule	operator	operator	operator	operator	operator	=	=	operator	operator	operator	operator	operator	operator		operator	habitat
	value	value	value	value	value	0	0	value	value	value	value	value	value		value	land only

IF airborne = 0 AND aquatic = 0 THEN habitat = land only

FIGURE 7.11 A habitat rule that applies to an aardvark.

			name	hair	feathers	eggs	milk	airborne	aquatic	backbone	breathes	fins	no of legs	tail	species	habitat	target	conclusion
run		case	aardvark	1	0	0	1	0	0	1	1	0	4	0			mammal	

order added	Go to if true	Go to if false	Rule no	name	hair	feathers	eggs	milk	airborne	aquatic	backbone	breathes	fins	no of legs	tail	species	habitat	target	conclusion
1	restart	2	1					= 1								mammal			
2	restart	exit	2						= 0	= 0							land only		

FIGURE 7.12 The rules screen after two rules have been added.

The conclusions for the two rules are outlined in black. If we click *run*, inference stops at the first conclusion and we get Figure 7.13 with *species* = *mammal* being the first conclusion.

			name	hair	feathers	eggs	milk	airborne	aquatic	backbone	breathes	fins	no of legs	tail	species	habitat	target	conclusion
run		case		1	0	0	1	0	0	1	1	0	4	0	mammal		mammal	

order added	Go to if true	Go to if false	Rule no	name	hair	feathers	eggs	milk	airborne	aquatic	backbone	breathes	fins	no of legs	tail	species	habitat	target	conclusion
1	restart	2	1					= 1								mammal			
2	restart	exit	2						= 0	= 0							land only		

FIGURE 7.13 The first inference step.

If we click *run* again, we get Figure 7.14.

			name	hair	feathers	eggs	milk	airborne	aquatic	backbone	breathes	fins	no of legs	tail	species	habitat	target	conclusion
run		case		1	0	0	1	0	0	1	1	0	4	0	mammal	land only	mammal	

order added	Go to if true	Go to if false	Rule no	name	hair	feathers	eggs	milk	airborne	aquatic	backbone	breathes	fins	no of legs	tail	species	habitat	target	conclusion
1	restart	2	1					= 1								mammal			
2	restart	exit	2						= 0	= 0							land only		

FIGURE 7.14 After clicking run a second time.

Rule 1 did not fire this time (coloured pink) because its conclusion species = mammal is already an asserted fact. If we click *run* again we get Figure 7.15.

				name	hair	feathers	eggs	milk	airborne	aquatic	backbone	breathes	fins	no of legs	tail	species	habitat	target	conclusion
		run	case		1	0	0	1	0	0	1	1	0	4	0	mammal	land only	mammal	
order added	Go to if true	Go to if false	Rule no	name	hair	feathers	eggs	milk	airborne	aquatic	backbone	breathes	fins	no of legs	tail	species	habitat	target	conclusion
1	restart	2	1					= 1								mammal			
2	restart	exit	2						= 0	= 0							land only		

FIGURE 7.15 After clicking run a third time.

Now both *species = mammal* and *habitat = land only* are asserted facts and neither rule 1 nor 2 fire for the case and the empty conclusion field is outlined.

Next we want to generate a final conclusion, and in this example, we want this final output to combine the *species* and *habitat* fields and output them in reasonable English. We would like an output such as

"The aardvark is of the species mammal and its habitat is land only".

Excel_RDR allows you to use VBA expressions to format conclusions. To give the desired conclusion, we enter the following expression in the conclusion field:

= "The" & K5 & "is of the species" & W5 & "and its habitat is" & X5 & "."

K5 is the rule builder worksheet cell for the *name* field for that case and similarly W5 and X5 are the cells for *species* and *habitat* for this case. To build the VBA expression above we can simply click on the values "aardvark", "mammal" and "land only" in the appropriate cells in the current case row and add the other VBA syntax as appropriate.

The system then puts "The aardvark is of the species mammal and its habitat is land only" in the conclusion field (Figure 7.16) and under it gives the VBA expression where the cell addresses are replaced by the variable names:

```
conclusion = "The" & name & "is of the species" &
species & "and its habitat is" & habitat & "."
```

It should be noted that the VBA expression must be constructed referring to the actual cells in the current case row (here, K5, W5 and X5); otherwise the expression will not correctly translate when the rule is run on the rule worksheet etc. This is simply a result of trying to develop a GRDR system in Excel.

We do not suggest this is a particularly useful output, Figure 7.16 merely demonstrates the use of a rule to construct an appropriate final output from the RDR. We want this rule to fire if there is both a conclusion for

backbone	breathes	fins	no of legs	tail	species	habitat	target	conclusion
1	1	0	4	0	mammal	land only	mammal	
1	1	0	4	0	mammal	land only	mammal	

backbone	breathes	fins	no of legs	tail	species	habitat	target	conclusion
operator	operator	operator	operator	operator	is_someth	is_someth	in operator	The aardvark is of the species mammal and its habitat is land only.
value	value	value	value	value			value	conclusion= "The " & name & " is of the species " & species & " and its habitat is " & habitat & "."

FIGURE 7.16 Constructing a rule to give a formatted conclusion.

species and *habitat* and so construct the following rule. The operator *is_something* is a Vba function defined in the module *user functions.*

If we accept this rule and go to the rules screen, on the third click of the *run* button we end up with the final conclusion as shown in Figure 7.17.

		name	hair	feathers	eggs	milk	airborne	aquatic	backbone	breathes	fins	no of legs	tail	species	habitat	target	conclusion
run	c case	aardvark	1	0	0	1	0	0	1	1	0	4	0	mammal	land only	mammal	The aardvark is of the species mammal and its habitat is land only.

order added	Go to if true	Go to if false	Rule no	name	hair	feathers	eggs	milk	airborne	aquatic	backbone	breathes	fins	no of legs	tail	species	habitat	target	conclusion
1	restart	2	1					= 1								mammal			
2	restart	3	2						= 0	= 0							land only		
3	exit	exit	3													is_something	is_something		The aardvark is of the species mammal and its habitat is land only.

FIGURE 7.17 The final conclusion given after run is clicked three times.

Note that since rule 3 is a final conclusion rule, it does not restart inference; inference goes either to a refinement rule changing the final conclusion or exits with the current conclusion as output.

Knowledge acquisition event: The next case that needs a rule is the bass. No rules fire so we add a rule for *species* in the same way as before. But this time we partially reduce the overgeneralisation used in SCRDR and MCRDR by using *aquatic = 1* and *backbone = 1* as rule conditions (Figure 7.18).

	name	hair	feathers	eggs	milk	airborne	aquatic	backbone	breathes	fins	no of legs	tail	species	habitat	target	conclusion
cornerstone																
current case	bass	0	0	1	0	0	1	1	0	1	0	1			fish	

	name	hair	feathers	eggs	milk	airborne	aquatic	backbone	breathes	fins	no of legs	tail		habitat	target	species
new rule	operator	operator	operator	operator	operator	operator	=	=	operator	operator	operator	operator		operator	operator	fish
	value	value	value	value	value	value	1	1	value	value	value	value		value	value	

IF aquatic= 1 AND backbone = 1 THEN species = fish

FIGURE 7.18 Adding a rule to specify species = fish for a bass.

We similarly add a rule to give the conclusion *habitat = water* with conditions *aquatic is 1* and *fins = 1.* If we then run inference stepping through the conclusions we end up with the final conclusion *"The bass is of the species fish and its habitat is water"*. This is correct but it is not particularly elegant, so we want to change this final conclusion. We add a rule to give the conclusion formatted as shown in Figure 7.19. The cornerstone case for the rule that gave the conclusion "*The bass is of the species fish and its habitat is water*" was the aardvark and is shown. We have to pick a condition that distinguishes the cases, so we use *habitat* = *water* and add a conclusion with slightly different wording.

backbone	*breathes*	*fins*	*no of legs*	*tail*	*species*	*habitat*	*target*	*conclusion*
1	1	0	4	0	mammal	land only	mammal	The aardvarkv is of the species mammal and its habitat is land only.
1	0	1	0	1	fish	water	fish	The bass is of the species fish and its habitat is water.

backbone	*breathes*	*fins*	*no of legs*	*tail*	*species*	*habitat*	*target*	conclusion
operator	*operator*	*operator*	*operator*	*operator*	*operator*	*is*	*operator*	**The bass is a fish lives in water.**
value	*value*	*value*	*value*	*value*	*value*	*water*	*value*	conclusion= "The " & name & " is a " & species & " lives in " & habitat & "."

FIGURE 7.19 A rule to improve the final output for a fish.

This results in the following set of rules. The rule trace shown is for the final conclusion for a bass. Note that both rules 5 and 6 have fired with rule 6, a refinement rule for rule 5 (Figure 7.20).

		run	case	*name*	*hair*	*feathers*	*eggs*	*milk*	*airborne*	*aquatic*	*backbone*	*breathes*	*fins*	*no of legs*	*tail*	*species*	*habitat*	*target*	*conclusion*
				bass	0	0	1	0	0	1	1	0	1	0	1	fish	water	fish	The bass is a fish and lives in water.

order added	Go to if true	Go to if false	Rule no	*name*	*hair*	*feathers*	*eggs*	*milk*	*airborne*	*aquatic*	*backbone*	*breathes*	*fins*	*no of legs*	*tail*	*species*	*habitat*	*target*	*conclusion*
1	restart	2	1					= 1								mammal			
2	restart	3	2						= 0	= 0							land only		
4	restart	4	3							= 1	= 1					fish			
5	restart	5	4							= 1							water		
3	6	exit	5													is_something	is_something		The bass is of the species fish and its habitat is water.
6	exit	exit	6														is water		The bass is a fish and lives in water.

FIGURE 7.20 The rules and final conclusion for a bass.

We then run the first four cases on the cases worksheet and get both the intermediate and final conclusions for each case – which are appropriate (Figure 7.21).

name	*hair*	*feathers*	*eggs*	*milk*	*airborne*	*aquatic*	*backbone*	*breathes*	*fins*	*no of legs*	*tail*	*species*	*habitat*	*target*	*conclusion*
aardvark	1	0	0	1	0	0	1	1	0	4	0	mammal	land only	mammal	The aardvark is of the species mammal and its habitat is land
antelope	1	0	0	1	0	0	1	1	0	4	1	mammal	land only	mammal	The antelope is of the species mammal and its habitat is land
bass	0	0	1	0	0	1	1	0	1	0	1	fish	water	fish	The bass is a fish lives in water.
bear	1	0	0	1	0	0	1	1	0	4	0	mammal	land only	mammal	The bear is of the species mammal and its habitat is land only.

FIGURE 7.21 Inference on the first four cases.

Knowledge acquisition event: The next case we need to deal with is the calf, since we earlier deleted the value for milk for calf. If we run the case for the calf we get the following rule trace (Figure 7.22).

	name	hair	feathers	eggs	milk	airborne	aquatic	backbone	breathes	fins	no of legs	tail	species	habitat	target	conclusion
run / case	calf	1	0	0		0	0	1	1	0	4	1		land only	mammal	

order added	Go to if true	Go to if false	Rule no	name	hair	feathers	eggs	milk	airborne	aquatic	backbone	breathes	fins	no of legs	tail	species	habitat	target	conclusion
1	restart	2	1					= 1								mammal			
2	restart	3	2						= 0	= 0							land only		
4	restart	4	3							= 1	= 1					fish			
5	restart	5	4							= 1							water		
3	6	exit	5													is_something	is_something		The calf is of the species and its habitat is land only.
6	exit	exit	6														is water		The calf is a lives in land only.

FIGURE 7.22 The rule trace for a calf after run is clicked for the first time.

The only rule that fires is rule 2 which gives *habitat* = *land* as its conclusion. The first rule for *mammal* could not fire because the value for milk was missing. In Excel_GRDR to assist users, we use a convention that if there is a missing value for a variable that could cause the rule to fire if it was known, then that rule is coloured purple. Even though *habitat* = *land only* has been asserted by rule 2 the final conclusion rules cannot fire as rule 5 requires a value for both *species* and *habitat*.

We could write a rule to classify a calf as a *mammal* based on other attributes, but we instead write a rule to get the *conclusion milk =1* as this will provide a demonstration of repeat inferencing. We guess the very overgeneralised rule shown in Figure 7.23.

	name	hair	feathers	eggs	milk	airborne	aquatic	backbone	breathes	fins	no of legs	tail	species	habitat	target	conclusion
cornerstone																
current case	calf	1	0	0		0	0	1	1	0	4	1		land only	mammal	
new rule	name	hair	feathers	eggs		airborne	aquatic	backbone	breathes	fins	no of legs	tail	species	habitat	target	milk
	operator	operator	operator	operator		operator	operator	=	=	operator	operator	operator	operator	operator	operator	1
	value	value	value	value		value	value	1	1	value	value	value	value	value	value	

IF backbone = 1 AND breathes = 1 THEN milk = 1

FIGURE 7.23 A rule to give the conclusion milk = 1.

The resulting knowledge base is shown in Figure 7.24. It shows the rule trace for the first click on the *run* button. Figures 7.24 to 7.27 show the rule trace after each further click on the *run* button.

	name	hair	feathers	eggs	milk	airborne	aquatic	backbone	breathes	fins	no of legs	tail	species	habitat	target	conclusion
run / case	calf	1	0	0		0	0	1	1	0	4	1		land only	mammal	

order added	Go to if true	Go to if false	Rule no	name	hair	feathers	eggs	milk	airborne	aquatic	backbone	breathes	fins	no of legs	tail	species	habitat	target	conclusion
1	restart	2	1					= 1								mammal			
2	restart	3	2						= 0	= 0							land only		
4	restart	4	3							= 1	= 1					fish			
5	restart	5	4							is 1			= 1				water		
7	restart	6	5					1			= 1	= 1							
3	7	exit	6													is_something	is_something		The calf is of the species and its habitat is land only.
6	exit	exit	7														is water		The calf is a lives in land only.

FIGURE 7.24 Rule trace when run is first clicked for the calf case, giving the conclusion *habitat* = *land only*.

				name	hair	feathers	eggs	milk	airborne	aquatic	backbone	breathes	fins	no of legs	tail	species	habitat	target	conclusion
		run	case	calf	1	0	0	1	0	0	1	1	0	4	1		land only	mammal	

order added	Go to if true	Go to if false	Rule no	name	hair	feathers	eggs	milk	airborne	aquatic	backbone	breathes	fins	no of legs	tail	species	habitat	target	conclusion
1	restart	2	1					=1								mammal			
2	restart	3	2						=0	=0							land only		
4	restart	4	3							=1	=1					fish			
5	restart	5	4							is 1			=1				water		
7	restart	6	5					1			=1	=1							
3	7	exit	6													is_something	is_something		The calf is of the species and its habitat is land only.
6	exit	exit	7														is water		The calf is a lives in land only.

FIGURE 7.25 Rule trace after run is clicked the second time for the calf case, giving the conclusion *milk = 1.*

				name	hair	feathers	eggs	milk	airborne	aquatic	backbone	breathes	fins	no of legs	tail	species	habitat	target	conclusion
		run	case	calf	1	0	0	1	0	0	1	1	0	4	1	mammal	land only	mammal	

order added	Go to if true	Go to if false	Rule no	name	hair	feathers	eggs	milk	airborne	aquatic	backbone	breathes	fins	no of legs	tail	species	habitat	target	conclusion
1	restart	2	1					=1								mammal			
2	restart	3	2						=0	=0							land only		
4	restart	4	3							=1	=1					fish			
5	restart	5	4							is 1			=1				water		
7	restart	6	5					1			=1	=1							
3	7	exit	6													is_something	is_something		The calf is of the species mammal and its habitat is land only.
6	exit	exit	7														is water		The calf is a mammal lives in land only.

FIGURE 7.26 Rule trace when run is clicked the third time for the calf case; *milk = 1* is now a fact so the first rule can now fire.

				name	hair	feathers	eggs	milk	airborne	aquatic	backbone	breathes	fins	no of legs	tail	species	habitat	target	conclusion
		run	case	calf	1	0	0	1	0	0	1	1	0	4	1	mammal	land only	mammal	The calf is of the species mammal and its habitat is land only.

order added	Go to if true	Go to if false	Rule no	name	hair	feathers	eggs	milk	airborne	aquatic	backbone	breathes	fins	no of legs	tail	species	habitat	target	conclusion
1	restart	2	1					=1								mammal			
2	restart	3	2						=0	=0							land only		
4	restart	4	3							=1	=1					fish			
5	restart	5	4							is 1			=1				water		
7	restart	6	5					1			=1	=1							
3	7	exit	6													is_something	is_something		The calf is of the species mammal and its habitat is land only.
6	exit	exit	7														is water		The calf is a mammal lives in land only.

FIGURE 7.27 Rule trace when run is clicked the fourth time giving the final conclusion.

The total sequence of evaluation is:
Rule 2 fires, *habitat = land only*, restart
Rule 5 fires, *milk = 1*, restart
Rule 1 fires, *species = mammal*, restart
Rule 6 fires, *final conclusion = "The calf is of the species mammal and its habitat is land only"*, exit

Rule 1 finally fired on the third inference cycle because *milk = 1* was asserted on the previous cycle.

Knowledge acquisition event: The next errors are for a chicken, so we need to add two rules to give a *species* and a *habitat* conclusion:

```
IF feathers = 1 THEN species = bird
IF airborne = 1 THEN; habitat = it lives on land
   and in the air
```

However, after adding these rules, the final conclusion is a mess as can be seen in the rule trace shown in Figure 7.28.

run		name	hair	feathers	eggs	milk	airborne	aquatic	backbone	breathes	fins	no of legs	tail	species	habitat	target	conclusion
	case	chicken	0	1	1	0	1	0	1	1	0	2	1	bird	it lives on lan	bird	The chicken is of the species bird and its habitat is it lives on land and in the air.

order added	Go to if true	Go to if false	Rule no	name	hair	feathers	eggs	milk	airborne	aquatic	backbone	breathes	fins	no of legs	tail	species	habitat	target	conclusion
1	restart	2	1					= 1								mammal			
2	restart	3	2						= 0	= 0							land only		
4	restart	4	3							= 1	= 1					fish			
5	restart	5	4							is 1			= 1				water		
7	restart	6	5					1			= 1	= 1							
8	restart	7	6			= 1										bird			
9	restart	8	7						= 1								it lives on land and in the air		
3	9	exit	8													is_something	is_something		The chicken is of the species bird and its habitat is it lives on land and in the air.
6	exit	exit	9														is water		The chicken is a bird lives in it lives on land and in the air.

FIGURE 7.28 Rule trace giving the final conclusion for a chicken.

The final conclusion is: "*The chicken is of the species bird and its habitat is it lives on land and in the air.*" In this case there is nothing wrong with the rules, but the way in which the conclusion is expressed is very clumsy. Rather than adding a rule to correct this we simply edit the *habitat* cell for rule 7. We enter *habitat* = *both land and air* and the final conclusion becomes "*the chicken is of the species bird and its habitat is both land and air*" as shown in Figure 7.29. This is not a particularly interesting change, but it makes the point that although one cannot change or edit the

run		name	hair	feathers	eggs	milk	airborne	aquatic	backbone	breathes	fins	no of legs	tail	species	habitat	target	conclusion
	case	chicken	0	1	1	0	1	0	1	1	0	2	1	bird	both land a	bird	The chicken is of the species bird and its habitat is both land and air.

order added	Go to if true	Go to if false	Rule no	name	hair	feathers	eggs	milk	airborne	aquatic	backbone	breathes	fins	no of legs	tail	species	habitat	target	conclusion
1	restart	2	1					= 1								mammal			
2	restart	3	2						= 0	= 0							land only		
4	restart	4	3							= 1	= 1					fish			
5	restart	5	4							is 1			= 1				water		
7	restart	6	5					1			= 1	= 1							
8	restart	7	6			= 1										bird			
9	restart	8	7						= 1								both land and air		
3	9	exit	8													is_something	is_something		The chicken is of the species bird and its habitat is both land and air.
6	exit	exit	9														is water		The chicken is a bird lives in both land and air.

FIGURE 7.29 A rule trace giving a better final conclusion for a chicken after the wording for the habitat conclusion was changed.

conditions in a rule in the RDR approach, the user is free at any time to change the way the conclusion concept is expressed. It is still the same rule, but the conclusion concept is expressed differently. However, as we noted earlier, whether it is wise to do this depends on the application, e.g. if a system rather than trying to deal with an individual case was assembling a collection of cases.

However, there is a second issue implicit in this change. With the SCRDR refinement structure used in Excel_GRDR there can be only a single habitat conclusion, so we cannot conclude both *land* and *air,* we have to give a single composite conclusion. In this regard we note that the earlier GRDR proposal (Compton, Kim, and Kang 2014) was based on repeat inference with a flat MCRDR rule structure where for any change of a conclusion a stopping rule and a new rule were added. This would have allowed both a *land* and an *air* conclusion to be made, but on the other hand this approach required all conclusions to be independent Booleans.

Knowledge acquisition event: The next case requiring a rule is the clam. It already has the *habitat* conclusion *land only* but there is no conclusion for *species.* We add the same rule for *mollusc* as for SCRDR (Figure 5.7) and for MCRDR. The final conclusion is then correct.

Knowledge acquisition event: The next case is for a crab. It is not misclassified as a *fish* as it was with SCRDR and MCRDR because we used a narrower rule for a *fish* (Figure 7.18), so we need to develop a new rule as shown in Figure 7.30.

	name	*hair*	*feathers*	*eggs*	*milk*	*airborne*	*aquatic*	*backbone*	*breathes*	*fins*	*no of legs*	*tail*	*species*	*habitat*	*target*	*conclusion*
cornerstone																
current case	crab	0	0	1	0	0	1	0	0	0	4	0			mollusc	
	name	*hair*	*feathers*	*eggs*	*milk*	*airborne*	*aquatic*	*backbone*	*breathes*	*fins*	*no of legs*	*tail*		*habitat*	*target*	*species*
new rule	*operator*	*operator*	*operator*	*operator*	*operator*	*operator*	=	=	*operator*	=	*operator*	*operator*		*operator*	*operator*	**mollusc**
	value	*value*	*value*	*value*	*value*	*value*	*1*	*0*	*value*	*0*	*value*	*value*		*value*	*value*	
	IF aquatic = 1 AND backbone = 0 AND fins = 0 THEN species = mollusc															

FIGURE 7.30 A new rule for an aquatic *mollusc*, rather than refining a *fish* rule.

We also add the rule:

```
IF aquatic = 1 THEN habitat = water
```

And the final conclusion is correct.

Knowledge acquisition event: The next problem case is that the system concludes a dove's *habitat* is *both land and air*, because of the rule

```
    IF airborne = 1 THEN habitat = both land and air
(Figure 7.28)
```

but it did not conclude a dove is a *bird* because we had deleted *feathers = 1*. We could write a rule to directly conclude a dove is a *bird*, but instead write the rule in Figure 7.31 to conclude *feathers = 1*.

	name	hair	feathers	eggs	milk	airborne	aquatic	backbone	breathes	fins	no of legs	tail	species	habitat	target	conclusion
cornerstone																
current case	dove	0		1	0	1	0	1	1	0	2	1		both land a	bird	
	name	*hair*		*eggs*	*milk*	*airborne*	*aquatic*	*backbone*	*breathes*	*fins*	*no of legs*	*tail*	*species*	*habitat*	*target*	*feathers*
new rule	*operator*	*operator*		*operator*	*operator*	=	*operator*	=	=	*operator*	*operator*	*operator*	*operator*	*operator*	*operator*	1
	value	*value*		*value*	*value*	1	*value*	1	1	*value*	*value*	*value*	*value*	*value*	*value*	

IF airborne = 1 AND backbone = 1 AND breathes = 1 THEN feathers = 1

FIGURE 7.31 A rule to conclude a dove has feathers.

After this rule is added the case is now given the correct conclusions and the inference sequence for a dove is:

- Rule 7 fires, *habitat = both land and air*, restart
- Rule 11 fires, *feathers = 1*, restart
- Rule 6 fires, *species = bird*, restart
- Rule 12 fires, *final conclusion = "The dove is of the species bird its habitat is both land and air"*, exit

Knowledge acquisition event: The next error is that a duck is misclassified as a *fish*. In adding a rule to correct the *fish* classification to *bird*, we can't use the *feathers = 1* conclusion from the rule we have just added in Figure 7.31, because this does not fire until after *fish* is concluded. We can only correct the conclusion *fish* by adding a refinement rule to rule 3 when rule 3 first fires – i.e. with the data available at that inference step. The rule we add, rule 4, is really a repeat of rule 12, the rule for concluding *feathers = 1*. The first inference pass is shown in Figure 7.32.

The need to add a repeat rule arises from RDR's use of linked rules so that *feathers = 1* is concluded after *species = fish* and the further constraint of not allowing a rule added after *feathers = 1* is concluded to change the earlier *species = fish* conclusion. This might seem like overkill in this very simple Zoo knowledge base, and perhaps these constraints could be loosened; however, it is precisely these constraints which allow for RDR's very rapid and simple knowledge acquisition, based on differentiating cases and avoiding the complexities and maintenance difficulties involved in trialing and testing changes to rules described in Chapter 1.

				name	hair	feathers	eggs	milk	airborne	aquatic	backbone	breathes	fins	no of legs	tail	species	habitat	target	conclusion
		run	c case	duck	0		1	0	1	1	1	1	0	2	1	bird		bird	
order added	**Go to if true**	**Go to if false**	**Rule no**	*name*	*hair*	*feathers*	*eggs*	*milk*	*airborne*	*aquatic*	*backbone*	*breathes*	*fins*	*no of legs*	*tail*	*species*	*habitat*	*target*	*conclusion*
1	restart	2	1					= 1								mammal			
2	restart	3	2						= 0	= 0							land only		
4	4	5	3							= 1	= 1					fish			
14	restart	restart	4						= 1		= 1	= 1				bird			
5	restart	6	5							is 1			= 1				water		
7	restart	7	6					1			= 1	= 1							
8	restart	8	7			= 1										bird			
9	restart	9	8						= 1								both land and air		
10	restart	10	9								= 0	= 0		= 0		mollusc			
11	restart	11	10							= 1	= 0		= 0			mollusc			
12	restart	12	11							= 1							water		
13	restart	13	12			1			= 1		= 1	= 1							
3	14	exit	13													is_something	is_something		The duck is of the species bird and its habitat is.
6	exit	exit	14														is water		The duck is a bird and lives in .

FIGURE 7.32 A refinement rule correcting *fish* to *bird*. The rule can't use *feathers = 1* since that is not asserted until later by rule 12.

When *run* is clicked again after adding rule 13 to conclude *bird*, the next rule that fires is rule 8 giving the *habitat* conclusion *both land and air.* We add a refinement rule to correct this to give a composite *habitat* conclusion including water (Figure 7.33).

	name	*hair*	*feathers*	*eggs*	*milk*	*airborne*	*aquatic*	*backbone*	*breathes*	*fins*	*no of legs*	*tail*	*species*	*habitat*	*target*	*conclusion*
cornerstone	chicken	0	1	1	0	1	0	1	1	0	2	1	bird	it lives on land and	bird	
current case	duck	0		1	0	1	1	1	1	0	2	1	bird	both land and air	bird	
	name	*hair*	*feathers*	*eggs*	*milk*	*airborne*	*aquatic*	*backbone*	*breathes*	*fins*	*no of legs*	*tail*	*species*		*target*	*habitat*
new rule	*operator*	*operator*	*operator*	*operator*	*operator*	=	=	*operator*	*operator*	*operator*	*operator*	*operator*	*operator*		*operator*	**water, air and land**
	value	*value*	*value*	*value*	*value*	1	1	*value*	*value*	*value*	*value*	*value*	*value*		*value*	

IF airborne = 1 AND aquatic = 1 THEN habitat = water, air and land

FIGURE 7.33 A refinement rule to correct the *habitat* conclusion for a duck.

After adding the rules, the first inference step is that rules 3 and 4 fire as shown in Figure 7.32 and the second step of inference is the firing of rule 8 and its refinement, rule 9 shown in Figure 7.34. In the inference

				name	hair	feathers	eggs	milk	airborne	aquatic	backbone	breathes	fins	no of legs	tail	species	habitat	target	conclusion
		run	case	duck	0		1	0	1	1	1	1	0	2	1	bird	water, air and land	bird	
order added	**Go to if true**	**Go to if false**	**Rule no**	*name*	*hair*	*feathers*	*eggs*	*milk*	*airborne*	*aquatic*	*backbone*	*breathes*	*fins*	*no of legs*	*tail*	*species*	*habitat*	*target*	*conclusion*
1	restart	2	1					= 1								mammal			
2	restart	3	2						= 0	= 0							land only		
4	4	5	3							= 1	= 1					fish			
14	restart	restart	4						= 1		= 1	= 1				bird			
5	restart	6	5							is 1			= 1				water		
7	restart	7	6					1			= 1	= 1							
8	restart	8	7			= 1										bird			
9	9	10	8						= 1								both land and air		
15	restart	restart	9						= 1	= 1							water, air and land		
10	restart	11	10								= 0	= 0		= 0		mollusc			
11	restart	12	11							= 1	= 0		= 0			mollusc			
12	restart	13	12							= 1							water		
13	restart	14	13			1			= 1		= 1	= 1							
3	15	exit	14													is_something	is_something		bird and its habitat is water, air and land.
6	exit	exit	15														is water		water, air and land.

FIGURE 7.34 The rule trace showing a refinement rule to give the correct habitat conclusion.

shown we have not yet got to rule 13 giving *feathers = 1*, but when this rule does fire it does not impact the final conclusion based on the conclusions from previous rules.

Knowledge acquisition event: The next error is for a flea; the *habitat* is correct for a flea, but no rule fires for *species*, so we add the same rule as in Figure 5.14, and the final conclusion is correct.

Knowledge acquisition event: Next a frog is classified as *species = fish*, with its *habitat = water* where it should be an *amphibian* living on land and in the water. As well, since the value for milk is missing for a frog, rule 6 asserted *milk = 1* since a frog has a backbone and breathes. We correct these three errors in the order in which they occur.

The first error is *species = fish*. We can't use the rule in Figure 5.15 which uses milk = 0 as a condition, because we have no value for milk at this stage, so we use *breathes = 1* and *no of legs > 0*. The next error is *milk = 1.* We add a simple refinement rule *IF eggs = 1* THEN *milk = 0* (Figure 7.35).

	name	*hair*	*feathers*	*eggs*	*milk*	*airborne*	*aquatic*	*backbone*	*breathes*	*fins*	*no of legs*	*tail*	*species*	*habitat*	*target*	*conclusion*
cornerstone	calf	1	0	0	1	0	0	1	1	0	4	1		land only	mammal	
current case	frog	0	0	1	1	0	1	1	1	0	4	0	amphibian		amphibian	
	name	*hair*	*feathers*	*eggs*		*airborne*	*aquatic*	*backbone*	*breathes*	*fins*	*no of legs*	*tail*	*species*	*habitat*	*target*	*milk*
new rule	*operator*	*operator*	*operator*	=											*operator*	0
	value	*value*	*value*	*1*		*value*	*value*	*value*	*value*	*value*	*value*	*value*	*value*	*value*	*value*	
	IF eggs = 1 THEN milk = 0															

FIGURE 7.35 A rule to correct a frog having *milk = 1* as a conclusion.

The final conclusion is "The frog is a amphibian and lives in water". This is probably reasonable, but is not grammatically correct (with "a amphibian") and we also need to change the *habitat*. We first add rule 15 which corrects rule 14 from giving the conclusion, *habitat = water* to habitat = *water and sometimes land* (Figure 7.36).

	name	*hair*	*feathers*	*eggs*	*milk*	*airborne*	*aquatic*	*backbone*	*breathes*	*fins*	*no of legs*	*tail*	*species*	*habitat*	*target*	*conclusion*
cornerstone	crab	0	0	1	0	0	1	0	0	0	4	0	mollusc	water	mollusc	
current case	frog	0	0	1	0	0	1	1	1	0	4	0	amphibian	water	amphibian	
	name	*hair*	*feathers*	*eggs*	*milk*	*airborne*	*aquatic*	*backbone*	*breathes*	*fins*	*no of legs*	*tail*	*species*		*target*	*habitat*
new rule	*operator*	*operator*	*operator*	*operator*	*operator*	*operator*	=	*operator*	=	*operator*	>	*operator*	*operator*		*operator*	water and sometimes land
	value	*value*	*value*	*value*	*value*	*value*	*1*	*value*	*1*	*value*	*0*	*value*	*value*		*value*	
	IF aquatic = 1 AND breathes = 1 AND no of legs > 0 THEN habitat = water and sometimes land															

FIGURE 7.36 A rule to change the habitat for an amphibian from water, to water and sometimes land.

The result of this change is that the final conclusion is given by rule 18, as rule 19 no longer fires. This gives the correct final conclusion "The frog is of the species amphibian and its habitat is water and sometimes land."

The sequence of rules shown in Figure 7.37 that fire on the frog case to reach the final conclusion is rules: 3, 5, 7, 8, 14, 15, 18.

				name	hair	feathers	eggs	milk	airborne	aquatic	backbone	breathes	fins	no of legs	tail	species	habitat	target	conclusion
		run	c case	frog	0	0	1	0	0	1	1	1	0	4	0	amphibian	water and	amphibian	The frog is of the species amphibian and its habitat is water and sometimes land.
order added	**Go to if true**	**Go to if false**	**Rule no**	*name*	*hair*	*feathers*	*eggs*	*milk*	*airborne*	*aquatic*	*backbone*	*breathes*	*fins*	*no of legs*	*tail*	*species*	*habitat*	*target*	*conclusion*
1	restart	2	1					= 1								mammal			
2	restart	3	2						= 0	= 0							land only		
4	4	6	3							= 1	= 1					fish			
14	restart	5	4						= 1		= 1	= 1				bird			
17	restart	restart	5									= 1		>= 4		amphibian			
5	restart	7	6							is 1			= 1				water		
7	8	9	7					1			= 1	= 1							
18	restart	restart	8				= 1	0											
8	restart	10	9			= 1										bird			
9	11	12	10						= 1								both land and air		
15	restart	restart	11						= 1	= 1							water, air and land		
10	restart	13	12								= 0	= 0		= 0		mollusc			
11	restart	14	13							= 1	= 0		= 0			mollusc			
12	15	16	14							= 1							water		
19	restart	restart	15							= 1		= 1		< 0			water and sometimes land		
13	restart	17	16			1			= 1		= 1	= 1							
16	restart	18	17				= 1				= 0	= 1				insect			
3	19	exit	18													is_something	is_something		The frog is of the species amphibian and its habitat is water and sometimes land.
6	exit	exit	19														is water		The frog is a amphibian and lives in water and sometimes land.

FIGURE 7.37 The rules to date, showing the inference path for the final conclusion.

Knowledge acquisition event: The next error is that a penguin is classified as an *amphibian*. The rule we added in Figure 7.32 doesn't apply because penguins do not fly. To change *amphibian* to *bird* we add a simple rule shown in Figure 7.38.

```
IF feathers = 1 THEN species = bird
```

	name	hair	feathers	eggs	milk	airborne	aquatic	backbone	breathes	fins	no of legs	tail	species	habitat	target	conclusion
cornerstone	frog	0	0	1		0	1	1	1	0	4	0	amphibian		amphibian	
current case	penguin	0	1	1	0	0	1	1	1	0	2	1	amphibian		bird	
	name	*hair*	*feathers*	*eggs*	*milk*	*airborne*	*aquatic*	*backbone*	*breathes*	*fins*	*no of legs*	*tail*		*habitat*	*target*	*species*
new rule	*operator*	*operator*	=	*operator*	*operator*	*operator*	*operator*	*operator*	*operator*	*operator*	*operator*	*operator*		*operator*	*operator*	bird
	value	*value*	1	*value*	*value*	*value*	*value*	*value*	*value*	*value*	*value*	*value*		*value*	*value*	
	IF feathers = 1 THEN species = bird															

FIGURE 7.38 A rule to correct *amphibian* to *bird*.

The final conclusion is then correct: "*The penguin is of the species bird its habitat is water and sometimes land*". But the wording is probably not ideal, so we edit the conclusion cell for rule 15 (Figure 7.37) and remove the word "sometimes". As before, changing the wording of a conclusion, does not change the rule or the concept being concluded – only its name. But note that this means the conclusion for a frog that was shown in Figure 7.37 will also change. There isn't a problem with this, it is still the same concept, but with a different label. The previous wording for the frog probably implied to anyone reading it that a frog spends more time in water than on land, while the wording for the penguin was changed because the person writing the rule did not want the conclusion to suggest the penguin was more in the water than on land. But in fact, the habitat rule that applied in both cases did not have a rule condition relating the relative amounts of time on water and land, so there is no problem, and it is probably more correct to change the wording. This is another example of the central problem in knowledge acquisition discussed in Chapter 2. One simply can never assume that a user's statement of a rule is some sort of universal general statement; it is always given in a context and the whole point of RDR is to be able to address these contextual limitations, as they are identified by relevant cases when they occur.

Knowledge acquisition event: The next error, as before, is that the pitviper does not have a *species* conclusion. We add the rule shown in Figure 7.39.

	name	*hair*	*feathers*	*eggs*	*milk*	*airborne*	*aquatic*	*backbone*	*breathes*	*fins*	*no of legs*	*tail*	*species*	*habitat*	*target*	*conclusion*
cornerstone																
current case	pitviper	0	0	1	0	0	0	1	1	0	0	1		land only	reptile	
	name	*hair*	*feathers*	*eggs*	*milk*	*airborne*	*aquatic*	*backbone*	*breathes*	*fins*	*no of legs*	*tail*		*habitat*	*target*	*species*
new rule	*operator*	*operator*	=	=	*operator*	*operator*	*operator*	=	=	*operator*	*operator*	*operator*		*operator*	*operator*	reptile
	value	*value*	*0*	*1*	*value*	*value*	*value*	*1*	*1*	*value*	*value*	*value*		*value*	*value*	

IF feathers = 0 AND eggs = 1 AND backbone = 1 AND breathes = 1 THEN species = reptile

FIGURE 7.39 A rule to classify a pitviper as a *reptile*.

This is perhaps a better rule than we used for SCRDR and MCRDR, and after adding it, the final conclusion is then correct.

Knowledge acquisition event: The next error is for a scorpion and adding the same rule as Figure 5.18 results in a correct final conclusion.

Knowledge acquisition event: As before, the next error is that a seasnake is classified as a *fish*. We correct it with the same rule as in Figure 5.19 and the final conclusion is then correct, although a little clumsy.

Knowledge acquisition event: The next error, as before, is that a slug is classified as an *insect*, and the same rule is added as in Figure 5.20 to correct this. The rest of the cases are correct and the final knowledge base is shown in Figure 7.40.

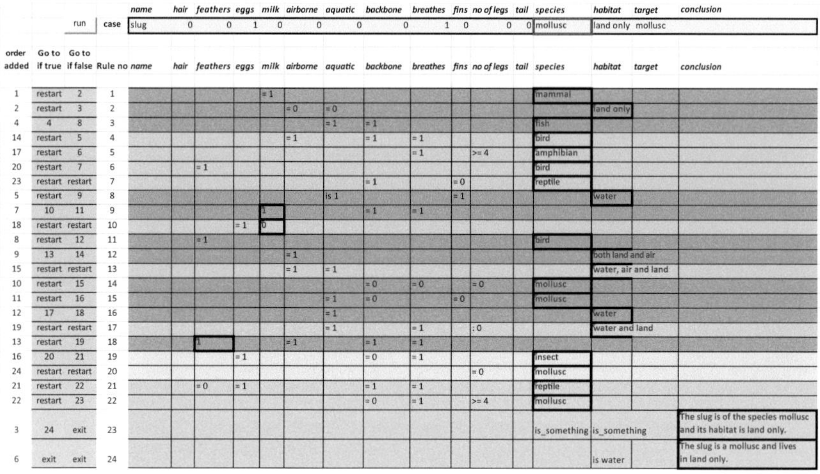

		name	hair	feathers	eggs	milk	airborne	aquatic	backbone	breathes	fins	no of legs	tail	species	habitat	target	conclusion
run	case	slug	0	0	1	0	0	0	0	1	0	0	0	mollusc	land only	mollusc	

order added	Go to if true	Go to if false	Rule no	name	hair	feathers	eggs	milk	airborne	aquatic	backbone	breathes	fins	no of legs	tail	species	habitat	target	conclusion
1	restart	2	1					= 1								mammal			
2	restart	3	2						= 0	= 0							land only		
4	4	8	3							= 1	= 1					fish			
14	restart	5	4						= 1		= 1	= 1				bird			
17	restart	6	5									= 1		>= 4		amphibian			
20	restart	7	6			= 1										bird			
23	restart	restart	7								= 1		= 0			reptile			
5	restart	9	8							is 1			= 1				water		
7	10	11	9					1			= 1	= 1							
18	restart	restart	10				= 1	0											
8	restart	12	11			= 1										bird			
9	13	14	12						= 1								both land and air		
15	restart	restart	13						= 1	= 1							water, air and land		
10	restart	15	14								= 0	= 0		= 0		mollusc			
11	restart	16	15							= 1	= 0		= 0			mollusc			
12	17	18	16							= 1							water		
19	restart	restart	17							= 1		= 1		: 0			water and land		
13	restart	19	18			1			= 1		= 1	= 1							
16	20	21	19				= 1				= 0	= 1				insect			
24	restart	restart	20											= 0		mollusc			
21	restart	22	21			= 0	= 1				= 1	= 1				reptile			
22	restart	23	22								= 0	= 1		>= 4		mollusc			
3	24	exit	23													is_something	is_something		The slug is of the species mollusc and its habitat is land only.
6	exit	exit	24														is water		The slug is a mollusc and lives in land only.

FIGURE 7.40 The completed knowledge base after processing all the Zoo dataset cases.

In Figure 7.40 it looks like we have a lot more rules than for SCRDR, but this is only because we have rules for five different types of conclusion; there are still only 13 rules for species, the same as Figure 5.22. What we have in Figure 7.40 is in effect five separate but interwoven knowledge bases for the five types of conclusion. After a conclusion is asserted for any one of the five conclusion types, inference returns to the start and rules giving conclusions of that type are no longer evaluated.

7.4 DISCUSSION: GRDR, MCRDR AND SCRDR

Our aim in including this chapter was to demonstrate a version of RDR that is as general as standard rule systems:

The key features of the Excel_GRDR system we have demonstrated are:

1. As with other RDR, rules are linked production rules.

2. During repeat inference cycles, "data" provided initially, or "facts" asserted by previous inference cycles, cannot be changed or retracted. Rules that conclude a value for an attribute which already has a value are simply not evaluated.

3. If an asserted "fact" is wrong, a refinement rule is added so that the wrong fact is not asserted in the first place.

4. GRDR inference returns to the first rule whenever a conclusion is asserted.

5. We also included a final output conclusion. The output conclusion can be refined exactly the same as other rules, but since it is the conclusion to be output, inference does not return to the first rule, and an output conclusion cannot be used in a rule condition – it has already been output and inference stopped.

We are not suggesting this is the only way to provide a general purpose RDR with looping, and what we have proposed here could be made more general. For example, stopping rules could be allowed. This would mean if inference passed down a refinement path and came to a stopping rule, the value of the variable being concluded would become null and inference would return to the false branch of the first rule that fired on the inference path. Inference would continue and some later rule might conclude a value for that variable, as it currently does not have a value. This would allow the user to use refinement rules, or at the other extreme only use stopping rules (which produce a composite rule with the parent) and a new rule when a correction was required. This would be the same structure we used for MCRDR, except the conclusions being inferred are all single-valued variables, so once a conclusion was asserted, no further conclusion for that variable could be asserted. A further way the current GRDR system could be generalised would be to allow variables that took set values. MCRDR and our earlier version of GRDR (Compton, Kim, and Kang 2014) allowed only a single set-value variable (the conclusion), but any number of variables could be allowed to take set values. Once a value was added to a set, inference would return to the first rule as there may be rule conditions relating to whether the set contained specific values.

We are not suggesting allowing for set conclusions as well as single value conclusions will have wide practical application, but if GRDR allowed for stopping rules and set conclusions as suggested, it would subsume both SCRDR and MCRDR. Both SCRDR and MCRDR have a single conclusion variable, except that for MCRDR this is a set variable. It would be up to the user to restrict (single value or set) conclusions to only a single variable, reproducing this restriction in SCRDR and MCRDR with GRDR.

A key practical issue for both GRDR and the more general version we have suggested is that the user may have to make ontological decisions, in deciding which variable a conclusion belongs to. In the example above when the value for *milk* or *feathers* was missing, it was obvious that the rule should assign a value to one of those variables. It was also pretty obvious that *species* conclusions were different from *habitat* conclusions. But it is not always this straightforward; it will depend on the application. For example, in interpreting blood chemistry results a pathologist in practice goes straight to the final composite piece of advice they want to send to the referring clinician. A domain expert like a chemical pathologist will happily identify ontological distinctions in such advice if requested, but as we discussed in Chapter 2, they will create these distinctions on the fly, depending on the context, and despite the distinctions being valid and insightful, they may do this slightly differently in another context. In terms of the dynamics of a human conversation, there are no issues at all in this, but for an expert system we expect the user to reuse the distinctions they have already made, unless they are clearly introducing a new concept. This means it is critical that a user is provided with interface support to prompt them about which variable they are providing a conclusion for, and also to reuse previous conclusions rather than constructing a different name for the same conclusion.

Apart from interface support to encourage the user to be consistent, perhaps there are other ways to minimise the problem. For example, perhaps when a refinement rule is being added (or a stopping rule plus new rule), the user could be queried as to whether the two conclusions could ever be given together. If not, they are probably alternative values for the same variable. As more refinements are made to values belonging to this variable, the question is repeated. If eventually the user says a refinement conclusion might possibly occur together with the conclusion it is replacing, a new variable would be added. On the other hand, if a rule with a new conclusion is added not as a refinement, and separate from the addition of a stopping rule, it would be considered as a new variable, but without this variable being named. If it was later refined / corrected by an existing conclusion, the user would be queried again, and this conclusion would be added to the possible values for the previous conclusion variable.

Perhaps set conclusions are a little easier to deal with. At the rule builder stage, the user could decide they wanted a rule to add another value to a variable that already had a value. If so these two values could be assumed to be members of a set. They could do this for any variable. The syntax available to the user for building rule conditions could be based on whether or not the value for the variable was a set. These simple measures would avoid the user having to make a priori decisions about what variables had set values. But although dealing with a set variable seems likely to be simple, it is probably of only occasional value.

We should emphasise that these ideas are pure speculation which have not been developed or tested but they do suggest that it might be possible to reduce the need of having to ask the user to name categories of conclusions. But interface support would still be required to encourage reuse of existing conclusion variables.

In this chapter we have demonstrated a GRDR system which we believe has the same generality as standard rule systems. A separate question is: what sorts of problems can be solved with such a system? SCRDR and MCRDR provide simple classification while GRDR obviously supports heuristic classification (Clancey 1985), where intermediate conclusions or abstractions are inferred before inferring a final conclusion from the intermediates. It also seems clear that GRDR should directly apply to configuration (parametric design) problems, where the solution is a set of variables whose values depend on each other, e.g. configuring DEC VAX computers (McDermott 1982, Soloway, Bachant, and Jensen 1987). A major focus of knowledge acquisition research has been on problem-solving methods (and related task methods) aiming to provide libraries of methods for solving different types of problems (Puppe 1993, Eriksson et al. 1995, Ten Teije et al. 1998, Motta 1999, Pérez and Benjamins 1999, Schreiber et al. 1999); however, what we have tried to present in this book is an alternative to machine learning, and machine learning is essentially used for classification. If a user wishes to apply RDR to other rarer problem types, it would be useful to consider some of this literature.

If a reader does wish to apply RDR to more complex problem types, we suggest a starting point would be to consider carefully the implications of the GRDR principles of not allowing facts to be retracted and having rules that construct the final output. In problems like resource allocation or perhaps text editing, the object the expert system is acting on is going to be changed, so what does it mean that facts cannot be retracted or changed? If an RDR system is developed for a problem like room allocation (e.g. Richards and Compton 1999) the RDR is going to assign people to rooms and perhaps move them, which looks like it is going to act on and change the data and conclusions it has previously made. If however, we consider an RDR system as having, even implicitly, both an input and an output, then the input that is passed to the RDR system is not changed within the RDR system, nor can a conclusion made by the RDR system be retracted within the RDR system. However, the output from the RDR system provided by an output rule, can be any sort of of recommendation or operator to change the data. But this change would take place outside the RDR system after which the RDR system may be called again to consider the changed data. What we are suggesting is that the problem-solving method that may be developed will be more than the RDR system – and also is more than any rule system. The RDR or other rule system takes in data and then passes some recommendation for action to the larger system in which it is embedded. In many applications it may seem as if the rules are directly changing the data, but at an abstract level, this is not actually the case, the result of inference is really an output string which is interpreted and acted on by the outside system.

Various industrial RDR projects have used various forms of repeat inference, since being able to use conclusions from rules as conditions in other rules can significantly reduce the knowledge acquisition effort as Clancey argued many years ago (Clancey, 1985). What we have tried to do in this chapter is to present some principles about repeat inference with RDR, but whether the particular approach we have demonstrated will produce better outcomes than more ad hoc methods, has not been demonstrated in practical application. Any repeat inference for heuristic classification or other more complex tasks will also make greater ontological demands on the domain expert, so we have also speculated how these may be reduced.

NOTE

1 Lindsay Peter, PKS chief technology officer, personal communication

CHAPTER 8

Implementation and Deployment of an RDR-Based System

In this chapter we consider some of the key practical issues with using an RDR knowledge-based system in an industry application. We start with the requirement for validation.

8.1 VALIDATION

The question is often raised: isn't it a problem that you have to keep monitoring the output of an RDR in order to catch any error cases (for which you then build rules)? The real question is the converse: under what circumstances would it be reasonable not to monitor the performance of a system to catch errors? Academic studies of machine learning have evolved excellent evaluation standards for comparing methods. Obviously, a knowledge base, developed by machine learning or any other method, must be evaluated on test cases, and these test cases should be unseen cases, not the cases used to build the system, and the gold standard method is 10-fold cross-validation as used in Chapter 5. Should systems be put into use based on these results without further monitoring?

RDR offers a different approach. Since rules are added over time whenever an error occurs, one can simply plot the number of rules added against the number of cases seen in adding those rules. The slope of the graph at any point will be an underestimate of the error at that stage. The graph in Figure 8.1 is from a repeat of the cross-validation studies described

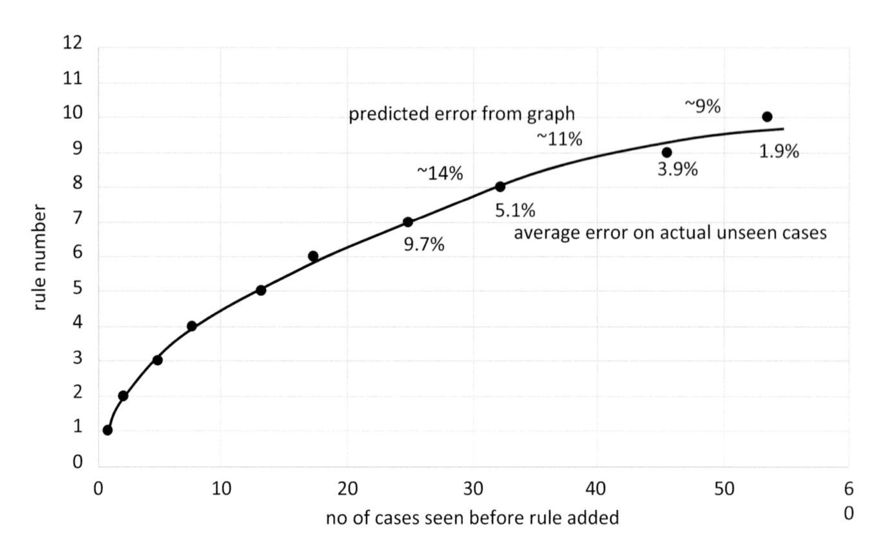

FIGURE 8.1 The number of rules added against cases seen and error rates for the cross-validation studies described in Chapter 5.

in Chapter 5. The X-axis is the average case number at which rules 1, 2, 3 etc. were added. There is of course considerable variability across the 10 studies, depending on whether or not there was a run of cases to which existing rules correctly applied, and whether rare cases occurred late or early, but only the average is shown here for simplicity. The numbers represent the percentage error at that stage. The percentages above the graph are rough calculations of the slope of the graph. For example, on average there were about seven correct cases from when rule 7 was added until another error resulted in the addition of rule 8, giving a very rough estimate of the error rate between rules 7 and 8 of ~14% (~1/7). The percentages below the graphs are the average actual errors on the remaining unseen cases in the 10 datasets after the rule was added. For each of the data points shown, the actual error is much less than the rough estimate of the error from the graph. This is simply because the estimated error is from before the rule was added, whereas the actual error on unseen cases includes the impact of the added rule. Obviously in a real industrial application, one would consider far more than 100 cases and would probably have far more than 10 rules before making estimates of errors. However, it is worth pointing out that in the 10-fold cross-validation studies we used 90 training cases in each study and found an error of 2% in unseen data whereas the representation of the data in Figure 8.1 shows that in fact an error of 2% was achieved after only 54 cases were seen, on average. An error of 10% is achieved after only 25 cases, no doubt because examples of more frequent classes are likely to occur early and because rules added to cover frequent cases greatly reduce the consequent error, but the data here is based on very rough estimates. The key point we wish to make is that RDR provides a very simple and conservative way of estimating errors, by looking at the rate at which errors have to be corrected.

The incremental validation and error assessment offered by RDR offers a challenge to conventional validation. If the only evaluation is pre-deployment, what is the likelihood, that the domain will change and the pattern of cases being processed by the system will change? David Spiegelhalter goes further and argues that accuracy is not the only thing that matters and that AI and machine learning should learn from the type of multi-stage testing used in introducing new drugs and medical protocols (Spiegelhalter 2020)[1].

And if a knowledge base is validated against a set of test cases, how good does it have to be: 95% accurate, 99% accurate, 100% accurate? Clearly, the performance required will change with the application. A stock market trading system may well be good enough with only 60% accuracy, or even less, as long as it is providing a good return on investment. On the other hand, a flight control system or an intensive care medical system ideally should not make an error. In reality systems often have a mix of accuracy requirements – some decisions are critical and should never be wrong but are others less mission critical. And what do you do if you find the system being validated has made a mistake? Can you fix it without having to revalidate the whole system?

We have raised these various questions to suggest that rather than the need for validation being a question for an RDR approach, a version of RDR style validation should probably be part of the gold standard for validating any system.

With an RDR approach, the user monitors the performance of the system and adds a rule whenever they want the system to produce a different output. As outlined in Appendix 1, industrial-use data suggests it takes only a minute or two to add a rule when required. The decision as to what level of performance is sufficient becomes a decision about the cost of ongoing monitoring versus the cost of an error being missed. It is also a decision entirely in the hands of the user or users, the people with the expertise doing the monitoring.

This case by case monitoring and relative cost approach allows for different performance standards for different conclusions and rules. The user can be provided with statistics for every case they consider: how often is a particular conclusion changed and a rule added, and how many cases have been given this conclusion since it was last changed etc.? Glenn Edwards coined the term "credentials" for such data, as we use these sort of "credentials" all the time in trusting human experts: how many heart transplants has Dr. X done? How many of the patients died? (Edwards et al. 1995). With this sort of credential information the user can decide:

- Which type of cases should always be referred for human validation because of either the particular conclusion reached, or the rule that fired?
- Which cases can be auto-validated, i.e. accepted without human checking or validation?
- What percentage of the cases with a given conclusion should be randomly referred for manual validation? The reasoning here is that the conclusion is likely to be correct, but the user wants to do some level of checking on whether new types of cases being given this conclusion are starting to emerge.

Laboratory customers of Pacific Knowledge Systems routinely use this sort of approach. Figure 8.2 shows the different levels of auto-validation across 185 knowledge bases from 12 laboratories. About 90 knowledge bases have very low levels of auto-validation with nearly all the cases being manually validated. It turns out that these 90 knowledge bases only process about 4% of the cases seen, so it seems perfectly sensible that these more rarely used knowledge bases, or perhaps knowledge bases at an early stage of their develeopment should be manually validated. In contrast 81% of cases are processed by knowledge bases with over 80% auto-validation. This suggests that if a lot of cases are seen it is likely the user will get to a stage where they are confident that few if any reports will need to be corrected. These are average figures; within each knowledge base the auto-validation rates are likely to be different across different conclusions and rules.

As we have already suggested, surely this detailed and ongoing validation which emerges naturally from the RDR approach, rather than being a burden, is probably the way all validation should be done.

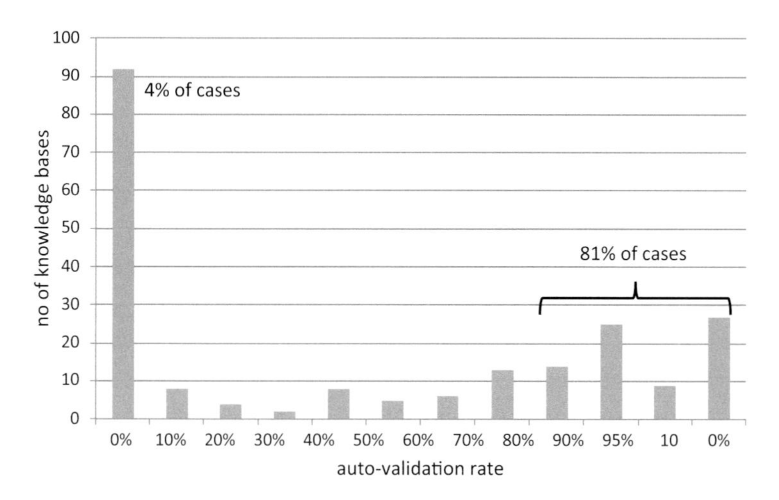

FIGURE 8.2 Auto-validation rates from logs of PKS data April to September 2011[2].

8.2 THE ROLE OF THE USER/EXPERT

From the early days of expert systems the role of the expert has been problematic. Did the expert have the time to invest in the period of intensive work required to develop the system? Would the expert be an involved and willing participant, or would they be reluctant? Would they see the project as appropriating their expertise; would they feel they might no longer be needed or as important?

The issues with RDR are quite different. The expert doesn't lose control of the project as there is no knowledge engineer intermediary; the rules are added by the expert and entirely at their discretion. It also doesn't take the expert away from their normal role of checking and reporting on data; rather rule addition is a very small addition to what they normally do, while at the same time enabling the output they would produce manually to be produced automatically, so they have the capacity to work on rules to produce even higher quality output, as well as other higher level tasks. Chemical pathologists who have written thousands of RDR rules well understand this advantage. They take the time to write rules that automatically produce consistent higher-quality patient-specific interpretative comments because the volume of cases being processed prevents them doing this manually. However, this also depends on the pathologist or other domain expert fully taking on the challenge. If the domain expert early on is assisted in learning how to write rules, but then fails to continue writing rules when they are on their own, an inferior RDR system will result. Although RDR avoids much of the problem of the domain expert's loss of ownership, it is still important to select a domain expert who will want to take ownership of the process of adding rules.

The central issue with RDR is not just that there needs to be stream of cases, but that the cases need to be monitored to pick up those cases where the right conclusion hasn't been made. Ideally this sort of monitoring should be part of someone's normal workflow rather than a new extra task; but if not, it has to be possible to arrange such monitoring, if an RDR approach is to be used. Ideally, the person monitoring the cases also writes the rules, but this is not strictly necessary. Cases might be monitored locally, but the knowledge base updated by the head-office expert, and we believe this is the case with some industrial RDR systems. In chemical pathology it is generally the chemical pathologist who authorises reports, and who adds rules as required. As noted in Appendix 1, Erudine used or proposed to use an RDR approach to re-engineer legacy software systems. The idea was that a case would be run through both the legacy system and the evolving RDR system and if they differed the business analyst responsible would add a rule or rules to the evolving RDR system. One would expect this would be a fairly short-term project until the new system was working well enough. Although this is an elegant idea and was advertised as a service, it is not clear whether Erudine actually used this approach on a significant project.

8.3 CORNERSTONE CASES

A central idea in RDR is that if a user decides a case has been given the wrong conclusion, they will be able to identify features in that case which distinguish it from cases for which the conclusion would be correct. So, what cases should be presented to the user, for them to select features that distinguish the new case with a different conclusion? Ideally all cases that had ever been processed by the old rule would be kept and all would be checked against the new rule, but this is not practical in a high-volume application. In the SCRDR example the case that prompted the addition of the previous rule was stored and the user was asked to identify features that differentiated the new case just from that case. No cases associated with other rules would reach the new rule, so this was very simple. The same approach was used with GRDR, but we have ignored a complication. With GRDR multiple conclusions might be made about a case, for example:

Rule A might conclude Conc1

Rule B might conclude Conc2

Rule C might conclude Conc3

Rule D might conclude Conc4

Rule E might conclude Conc5

Now if conclusion Conc4 is wrong the user will add a refinement rule D1 to give conclusion Conc4a, and the cornerstone case for adding the new rule is the case associated with Rule D and conclusion Conc4. The rule is added, and the case is stored as a cornerstone case for Rule D1. The problem with this is that the cornerstone cases for both Rule D and Rule D1, also fired Rules A, B and C on the way to Rule D. Now if one of these rules was later refined, say Rule B, only its cornerstone would be considered, not the cornerstones for Rules D and D1, which also fire Rule B. This means a change might be introduced which wasn't checked against these cornerstones and they might now be misclassified. And of course, if Rules D and D1 also used Conc2 from Rule B as a rule condition, then the problem is magnified.

Although this problem is real, we believe it is not a major issue, and a study of errors on past correct cases, caused by new rules indicated errors were likely to be 2% or less, even early in the development of the knowledge base (Kang, Gambetta, and Compton 1996). This was a study using a simulated expert so it is likely a real expert would do better in creating rules.

One way to address this problem with GRDR is that a case for which a new rule is added becomes a cornerstone case for every rule it fired, with the case stored with each rule with whatever conclusions had been asserted before that rule was evaluated. The other approach is simply to save every case for which a rule is added and check any new rule against them all, one be one. The same issue arises with MCRDR in that multiple conclusions may be made by multiple rules. In the version of MCRDR we demonstrated, refinement rules giving a new conclusion were not used, rather we used stopping rules to prevent a conclusion being given and added a new rule to give the right conclusion. Because new rules are added at the top level, every cornerstone case might fire the new rule, and the worked example showed them all being checked and the user gradually adding sufficient conditions to exclude all the cornerstone cases. From PKS experience it seems that a domain expert only has to see two or three such cases out of perhaps thousands before they have developed a sufficiently precise rule to exclude all these cases. If refinement rules are used with MCRDR then the case for the parent rule needs to be checked against the new rule, as well as cases associated with any other refinements of the parent rule. There is also the same issue as with GRDR in that other rules might fire for this case, but the case will not be stored as a cornerstone case for these rules, it will only be stored for the refinement rule added.

This probably sounds a bit messy, but we have already suggested the simple solution that whenever a rule is added, all the cases stored with all the other rules are checked. As suggested, domain experts seem to be able to make a sufficiently precise rule to exclude all cornerstone cases after seeing at most a few cases, shown one by one. As extra conditions are added to the rule more of the cases waiting to be checked are excluded. The issue with this approach is that the inference needs to run fast enough so the user experiences no delay in waiting to see what cases, if any need to be considered. Clearly some sort of data base retrieval can also be used to find candidate cases, or cases could be presented in some sort of order, but the key point is that in reality, domain experts, because they are experts, need to see very few cornerstone cases before they have constructed a rule that excludes all cornerstone cases.

8.4 EXPLANATION

In theory, no explanation is needed with an RDR system; the user sees a case for which they want to change or add a conclusion, and when changing a conclusion, they are simply asked to identify features that distinguish that case from other cases. That is the theory, but in practice users are likely to want to know why a particular conclusion was given. This is particularly the case when the conclusions the system provides are subtle. In a conventional rule-based system, the rules are likely to be organised in some structured and meaningful way, at least when the system is first deployed, so browsing the knowledge base can be helpful. This isn't helpful with RDR rules as the knowledge-base structure is based on when different cases occurred.

With an RDR system obviously you can get a rule trace similar to other knowledge-based systems. This is helpful, but the biggest advantage of the RDR approach is that each of these rules has a case associated with it, so the user can see the original cases for which the rules were made, and which have resulted in the particular conclusions for the case at hand. Case-based reasoning (CBR) has long argued reference to cases is central to much human problem solving (Schank and Abelson 1977, Kolodner 2014). RDR does not provide the central feature of CBR of adapting a case to provide a solution for a new situation, but it can provide the cornerstone cases, in all their detail, for all the rules which apply to the case in hand, even those which were stopped or refined, enabling the user to see the full context for which the rule was constructed. If a developer wishes to go further and provide some sort of way to browse all the knowledge in an RDR system, they could provide tools for users to construct and modify cases to test against the rules, but a more comprehensive solution would be to use something like a Formal Concept Analysis lattice for the user to browse (Ganter and Wille 2012). Essentially an RDR rule or rule path becomes an object and the rule conditions the attributes of these objects resulting in a knowledge-base lattice that a user can browse (Richards and Compton 1997)

8.5 IMPLEMENTATION ISSUES

Traditionally an organisation wanting to develop a knowledge-based system would either obtain a so-called expert system shell, or use an information system with an embedded expert system component. For example, IBM bought ILOG and now provides IBM WebSphere ILOG JRules. The reason such products are attractive is that as well as integration with a supplier's other systems, such tools have sophisticated high performance inference engines as well as rule development environments, giving graphic representations of the knowledge base etc.

RDR systems are completely different. With RDR there is no inference engine or rather a trivially simple engine. The knowledge base itself specifies what rule is to be evaluated next so there is no need for an inference engine's conflict resolution strategy to determine which rule to fire. Because of this structure it is also trivial to compile the rules into a procedural program. The main interface effort in developing an RDR system is not a system to browse the knowledge base; the key challenge is developing ways of representing cases and identifying features that will be used in rule conditions appropriate for the domain and easy for a domain expert to use. But this is something that has to be done for each different kind of project, regardless of whether you are using RDR or any other method, if you are going to produce a system that is reasonably easy for the domain expert to use. This development will generally be on a project by project basis, except where the same type of data is processed in the same way by different organisations. For example, most PKS customers have very similar requirements for case and feature representation, so it makes sense for them to use the software provided by PKS.

This leads to the suggestion that unless there is an RDR implementation available with an interface directly designed for the target application, it is probably simpler for an organisation to build their own RDR system, rather than trying to adapt some sort of general RDR tool – e.g. the very simple Excel_RDR systems described here. For example, having a separate column for each attribute as in Excel_RDR would get very clumsy with a large number of attributes, and cannot be used for time course data. The best approach, if feasible, is to use the same representation of a case that users are already familiar with. Then on top of this, one can use mouse-over techniques etc. to show the user what sort of rule conditions they might be able to construct from a particular attribute.

In summary, a critical issue for RDR, and for any other expert system, is to provide an interface that is appropriate for the domain and so is easy to use. This may require a significant development project, but the trade-off with RDR is that it is comparatively very simple to write an RDR rule engine. However, there is also an issue in how quickly the system needs to process a case. In some applications extremely high performance will be required, requiring not only coding a high-performance program but also consideration of data bases and the location of the system in a network. But these are all standard issues to be addressed with any system.

8.6 INFORMATION SYSTEM INTERFACES

Stand-alone expert systems are virtually useless. In what circumstances will a user want to manually enter data into a system to then get advice on what the data means? People will do this on a one-off basis, but to deal with a stream of cases in this way will nearly always be a waste of everyone's time. An exception to this was the Oncocin system (Shortliffe et al. 1981). Apparently, staff would come and enter the oncology data for a patient, but then some would walk away before Oncocin came back with its advice. It seems that staff liked to use it, because it organised and presented data very clearly – so clearly, that they were able to reach a conclusion about the data before Oncocin came to its own conclusion (Bruce Buchanan, personal communication).

In general, an RDR or any other knowledge-based system will need to be interfaced to an existing information system, obviously to provide the stream of cases and also to take the output from the RDR system. If an external company is providing or developing the RDR system, there are issues with the level of access between the RDR system and the information system. Can the RDR system query the information system and post data to it or does it have to wait until it is passed data and then advise the information system that it has processed the data? Once these policy issues are resolved there is the further issue with the actual interface. Customers of an RDR provider might have quite different system interfaces, although they handle very similar types of cases requiring the same user interfaces. Increasingly, web services or similar would be used to deploy an RDR system. The point of this discussion is to emphasise that although RDR and an RDR inference engine are extremely simple ideas, the challenge in deploying an RDR, or any other system, is in providing the appropriate interfaces. Both of us in supervising RDR PhD projects have expected students to develop their own RDR engine. Not only do they do this quickly, but the student is then in complete control of the project, and is more easily able to try out variations of RDR for the problem they are working on and develop interfaces etc.

NOTES

1 https://www.facebook.com/nipsfoundation/videos/482957018893956/
2 This figure first appeared in a PKS white paper "Auto-validation with RippleDown" by P.Compton https://pks.com.au/technology/resources/white-papers/

CHAPTER 9

RDR and Machine Learning

This book started with a discussion of machine learning and suggested that obtaining sufficient suitable data for machine learning could be a problem, and the promise of machine learning difficult to achieve. It was then argued that the alternative was not conventional expert systems with their separation of knowledge and inference, but Ripple-Down Rules (RDR). We worked through detailed examples of various versions of RDR to try to make clear exactly how RDR works, and followed this with a discussion of implementation issues with real-world systems. Hopefully RDR is now sufficiently clear that we can revisit machine learning to look at how RDR relates to machine learning and how it can complement it.

RDR is much more closely related to machine learning than most knowledge acquisition methods although there are some methods like repertory grids based on Kelly's Personal Construct Psychology that are also based on users differentiating cases (Gaines and Shaw 1993). Machine learning depends on a set of appropriate cases and the machine learning algorithm, whether decision tree, some sort of deep network learner, support vector machine or whatever, always builds a particular structure and applies a particular statistical strategy to try and figure out the relative importance of the different features in reaching different conclusions. RDR also uses a fixed structure, with the structure varying with the type of RDR used. In RDR the statistical algorithms are replaced by a user identifying the features that distinguish cases with different conclusions, and this is done on a case by case basis. Because of this structural similarity with machine learning, a number of machine learning algorithms have been developed that produce an RDR representation and these will be discussed later in this chapter. There is also some work trying to bring together case-based RDR and more formal representation methods (e.g. Gaines 1991b).

In contrast, other knowledge acquisition methods are more about a software engineering approach to knowledge engineering. CommonKADS, probably the flagship for such approaches considers the overall development process for a knowledge-based system from understanding what the task and context is, to selecting an appropriate problem-solving (or task) method, building a model, knowledge elicitation, etc. (Schreiber et al. 1999). This is a powerful approach for understanding a new problem and domain and no doubt this type of software engineering thinking would also be useful in starting work on a machine learning project. RDR like most machine learning projects has the much narrower aim of dealing with a pre-existing stream or data base of cases and making decisions about the data in future cases. Before looking more closely at the relationship between machine learning and RDR, we will revisit the issue of a suitable stream of cases.

9.1 SUITABLE DATASETS

Given the similarities between RDR and machine learning, when should RDR be used rather than machine learning? Obviously, machine learning should be used instead of manual knowledge acquisition if there are large well-curated datasets with reliable labelling of data. But as was discussed in Chapter 1, how often does this occur in real-world application?

Pacific Knowledge Systems provides RDR technology mainly for chemical pathology (clinical chemistry). There are at least 800 PKS user-developed RDR knowledge bases in use ranging from 100s to over 10,000 rules (Appendix 1) and some of these produce 100s of different conclusions. These can be combined (MCRDR is used) resulting in a knowledge base giving perhaps 1000s of different pieces of advice. The reason for such fine-grained conclusions is that the goal for pathology practices is to provide the detailed patient-specific advice that a pathologist might provide to a GP, if they had the time to call the GP or write a detailed comment. There will be subtle differences in this advice depending not just on current results, but past results, clinical history, age, medications and even who the referring clinician is, and why they have requested various tests. Machine learning applied to medical data would probably aim for learning coarser diagnostic categories, and it could be argued that this is all that necessary for diagnosis and management. But even so, is suitable data available? On the other hand, with RDR the domain expert can assign whatever conclusion they want to the particular case for which they are adding a rule; it doesn't matter how subtle it is, except that more subtle conclusions will probably require more rules.

Rashidi et al. note that medical data is growing at the rate of 48% per year and the majority of the data in medical records is from the clinical laboratory, so it seems like a very fertile ground for machine learning (Rashidi et al. 2019). But they go on to comment on the difficulties of getting suitable data:

- Test methods can lack standardisation so that results from one lab, for say cardiac troponin, may not be comparable to another due to differences in the method.
- Medical data are often incomplete, which is a challenge for learning.
- This leads to investigators using more complete and rigorous data from clinical trials, but such data may not represent the real-world population and may result in overfitting.

They do not comment on the issue of accurate labelling, which as discussed in Chapter 1 is a problem even for the relatively simple discharge coding often used for determining funding. The important point for RDR is that none of these limitations of the data are a significant problem. As long as you have a domain expert able to provide interpretations of whatever data is available and justify why a different interpretation is given for different data, you can go ahead with an RDR development.

A 2017 paper entitled "Machine learning in laboratory medicine: waiting for the flood" (Cabitza and Banfi 2018) notes a 10-fold increase in papers about machine learning in medicine, but in looking for papers specifically on laboratory medicine they go on to comment:

> Finding so few articles on the application of ML to traditional laboratory parameters was somewhat unexpected, as the laboratory is regarded as the main supplier of quantitative, structured and codified data in clinical medicine.

They report on various actual machine learning studies in chemical pathology, but these seem to be careful studies in specific areas with well-worked-up data, rather than reports on the significant use of machine learning knowledge bases in clinical practice. Many of the papers on machine learning in chemical pathology seem to focus on the promise and possibilities for machine learning in chemical pathology, expressing similar sentiments to those of Cabitza and Banfi (2018) despite so few articles to date on actual use:

> Nonetheless, we expect that ML methods will become more extensively used in the analysis of laboratory parameters, and especially for data that can be easily grouped and compared across different groups.

Although not referring to laboratory data, IBM's 2012 Global Technology Outlook provided a warning about the expectations arising from increasing data availability. It started with:

> Analytics provides new opportunities for business insight through the analysis of data that is constantly growing in volume, variety, velocity and uncertainty. Increases associated with the first three categories have been evident and widely acknowledged, but decreases in veracity represent the fastest-growing portion of data that is uncertain.
>
> (IBM Research 2012)

We note once again, that with an RDR approach, as long as you have data and a domain expert you can start to develop a system. In fact, if the domain expert who would develop the rules is also the person checking lab reports as they are issued, then the RDR development fits into the natural workflow of the laboratory.

9.2 HUMAN EXPERIENCE VERSUS STATISTICS

We have already presented the trivial example of the Zoo dataset where the RDR system significantly outperformed the J48 decision tree learner, using the same 10-fold cross-validation evaluation. No doubt machine learning would have done much better with more training examples – but on the other hand, as we have previously noted there were data errors even in 101 cases available in a dataset from the UCIrvine data repository, so there has to be a question of whether a larger dataset would be of good enough quality.

A better example of people having useful background knowledge is the IBM project on data cleansing for Indian street address data (Dani et al.

2010). As discussed in Chapter 1, all the methods, except for the commercial system, perform similarly when tested on data similar to the Mumbai data on which they were trained, with a precision of 75–80%. However, when tested on data from all of India, although all methods degrade, the RDR method degrades much less than the other methods. A more sophisticated RDR method is being used than a single SCRDR knowledge base, but it is clear that the general background knowledge of a human user produces a more generally applicable knowledge base than can be achieved by purely statistical methods in this sort of domain.

As we discussed in Chapter 1 a critical question for machine learning is the origin of the labels needed for supervised learning. The machine learning methods used in the Indian street address study would have done better with more training data from other parts of India. But such training data needs to be labelled by people; which in this case means a correct address has to be matched with an ill-formed address. If this could have been automated there would have been no need for this project, which as discussed in Appendix 1 has resulted in a commercial deployment. If people have to do the matching, they can also write rules as they go, and if the PKS data is typical, this will only take them a couple of minutes. But we should also note the research study described below, where 7000 emails were checked and 372 rules written to classify them in 8–9 hours (Wobcke, Krzywicki, and Chan 2008). Including checking the 7,000 cases, this comes out at less than 90 secs per rule; the paper states that actual rule building was about 25 secs per rule. This was a research study but taken together with the PKS experience; it is obvious that rule building is extremely rapid – because the user simply has to identify features that differentiate cases to which they wish to assign a different label. So, in trying to develop a good enough dataset for machine learning, one can very rapidly build an RDR system to do the same task, and as the simple Zoo example and the Indian address study suggests, probably with far fewer cases than the number needed for machine learning to work as well as the RDR. Will there be less human effort required overall in building an RDR knowledge base than in labelling sufficient cases for a machine learner to produce a knowledge base of the same quality?

9.3 UNBALANCED DATA

A related issue is the question of unbalanced training data. Even if the data is correctly labelled can a learner successfully learn to recognise rare events? There are many papers on different ways of trying to deal with unbalanced datasets and the methods proposed vary with the machine learning method. A widely cited 2017 review found 517 papers on how to deal with unbalanced data in the previous decade (Haixiang et al. 2017). The review notes that the problems with unbalanced data include:

- Learners like decision trees and support vector machines tend to be biased towards a good coverage of the majority examples.
- Performance metrics such as prediction accuracy are also biased towards the majority.
- Pruning, intended to get rid of noise in the data may also get rid of rare examples.
- Minority examples usually overlap with other regions where the prior probabilities of both classes are almost equal.
- Small sample sizes together with many attributes exacerbate the difficulty in learning the minority class or classes.

Methods to try to deal with unbalanced data include resampling where either new minority class examples are created, or some majority class examples are removed. Another approach is to reduce the number of features. Also, rather than simple overall accuracy, errors in the minority class can be given a greater cost to improve learning. Ensemble methods are also used where a combination of classifiers is better than a single classifier.

The point of this discussion is not to discuss the machine learning problems with unbalanced data, but simply to suggest that this problem does not arise with RDR. A domain expert adding an RDR rule adds the rule for a single case regardless of whether or not it belongs to a majority class or one of the rarer classes. The problem with unbalanced data only occurs when one is trying to use some sort of statistical method. Of course there is still an apparent problem with RDR in that rare cases will occur rarely and so it may take a long time for these rarer cases to show up in the stream of cases, perhaps requiring almost perpetual monitoring of cases and seeming to nullify the apparent advantage of an expert being able to deal with a single case.

This apparent limitation resolves when we consider where the labels for the training data might come from. Firstly, if the labels are produced from some definitive outcome, e.g. failure to repay a loan, then it is likely that every case has a label which we can assume is correct. If a machine learner is used to build a knowledge base with this data, it doesn't really matter if the resulting knowledge base is biased because of unbalanced data, because the training cases where it does not give the right answer can be passed to the expert who can construct rules for these cases on a case by case basis. This idea of an RDR wrapper to correct the output of another system is discussed further below. On the other hand, if the cases are labelled by the domain expert, they will probably have to see as many or more cases to ensure they have labelled at least some examples of rare cases, than if they were building rules at the same time. As we have noted, industrial experience suggests that writing a rule should only take a minute or two on average, so it is perfectly feasible for the expert to write rules at the same time as they label cases. It seems fair to say that RDR can provide an effective solution for the problem of unbalanced data which might be preferable to manipulating the data by generating or removing cases or trying to investigate feature relevance.

A related issue is that the distribution of cases available for learning may change over time. This is an obvious issue with problems like classifying email, as the distribution of emails will tend to change as topics emerge and fade over time. However, this issue also arises in less obvious situations and was investigated using 43,000 sets of thyroid clinical chemistry results from 1979 to 1990 (Gaines and Compton 1995). Brian Gaines used his Induct-RDR machine learner (described below) to learn from eight sequential sets of data of about 5,000 cases each and then tested the resulting knowledge bases on each of the other datasets. In general, when the test data was close in time to the training data the error rates were a few percent, but when the most recent sets of data were run as test data on knowledge bases trained on early datasets the errors were 20%–30%. Conversely when early test data was run on knowledge bases learned from the most recent datasets, the error rate was 10%–15%. Models were also trained on the first 500 cases, then the first 1000, then the first 1500 and so on and then tested on the following 500 cases, as this probably resembles what might occur in practice, that you train a learner on past cases and then start running it on new cases. In general, the error was low except for two occasions when it jumped up to about 12% on the next 500 cases. This is a challenging dataset as there are over 50 classes, and the data is also very unbalanced with the majority class (i.e. no conclusion implying no thyroid disease) being over 70% of the data; however, clearly the data and/or its distribution of classes changes over time. Perhaps the changes were because of different analytical tests used over the years, although it was assumed pre-processing corrected for this. Secondly, probably the specialists and registrars doing most of the test ordering also changed over time and perhaps the tests they ordered and the population of patients they treated may also have changed. We do not know exactly if or why the population might have changed, but what this dataset does make clear is that selecting appropriate training data is not as straightforward as it might seem. With an RDR system built at the time, we can probably assume the domain expert is aware of changes in analytical techniques and who is ordering tests and why. They can then simply add rules if and when necessary to give the output appropriate at the time.

But this again raises a central question with RDR: that developing an RDR system is based on monitoring the RDR system's output and adding rules as required. We have already discussed validation with RDR and how the domain expert can decide some classes (or even different rules for the same class) may no longer need to be monitored, or only monitored occasionally to check if there are any changes coming through. We have also argued that this in fact is a superior approach to initial evaluation on test data, and then trusting that everything will be OK. We have further argued that if one is trying to label cases for machine learning, it is probably more efficient to write rules at the same time – and labelling is likely to be more consistent, as different labels have to be justified in terms of differentiating features. We could also argue that monitoring to check that labels generated by an evolving RDR system are appropriate is an easier task than having to check cases without labels and assign a label or classification. Despite all of this, it would be preferable if the amount of monitoring could be reduced in a more reliable way than the expert simply deciding that since rule X always gives an answer they accept, that it needs to be monitored only very rarely. One way to approach this issue is through the notion of prudence.

9.4 PRUDENCE

This discussion of rare cases raises the question of whether an RDR system might be able to detect and warn whether it is seeing a particular type of case for the first time. Early expert system literature referred to the "brittleness" of expert systems (Lenat, Prakash, and Shepherd 1985); that is, an expert system may make a mistake that even a novice human expert would avoid. A junior intern might not know what is up with the patient, but they know they should refer the case to someone with more expertise and experience. What the novice is doing is recognising that there is something about the case which is outside their range of expertise. Can RDR do the same?

There has been a range of research with RDR trying to find ways of recognising that the conclusion for a case might be wrong, and that this case should be referred to a human expert for rule building. This should be a goal for all knowledge-based systems, but attempting to achieve this arises naturally with an RDR approach. This type of work commenced with the credentials idea outlined above (Edwards et al. 1995) and included work on trying to recognise whether a case differed from previously seen cases (Compton et al. 1996). There has been a wide range of this type of work including Prayote 2007, Prayote and Compton 2006, Finlayson and Compton 2013, 2014, Maruatona, Vamplew, and Dazeley 2012a, Dazeley, Park, and Kang 2011, Dazeley and Kang 2008, Maruatona et al. 2017, Amin et al. 2018, Amin et al. 2017 and Amin et al. 2015. The simplest idea is just to look at whether attribute values of those cases which followed a particular rule path are outside the range of previously seen attribute values for such cases. Dazeley used a neural network to look at the output of an MCRDR system to decide which of the conclusions given for a case were doubtful (Dazeley, Park, and Kang 2011). This approach seemed particularly useful with a document domain where many independent conclusions may be given. Much of the work on prudence has focused on specific domains of interest. A more general-purpose method, evaluated on a range of standard machine learning datasets was a variant of semi-supervised learning (Finlayson and Compton 2014). Cases were not only labelled by the user, but a rule was also added. Some initial cases were labelled, and rules were added, but once there were enough cases for machine learning, further rules were added (by a simulated expert) only when an evolving machine learning system gave conflicting advice to an evolving RDR system. Evaluation on a range of standard datasets suggested this approach outperformed other semi-supervised learning (Finlayson and Compton 2014).

Unfortunately, none of this research has achieved the goal of identifying all cases where the knowledge-based system (KBS) is about to make a mistake, so these techniques cannot be used to eliminate the need for some level of human monitoring. Secondly, there is the classic false-negative, false-positive trade-off. The fewer the false-negatives (missed errors) the greater the false-positives – correct cases unnecessarily referred for manual checking. We believe that these approaches, although not perfect could perhaps provide a useful aid in monitoring the output of a KBS, whether RDR or any other type of KBS. If, for example, 10% of cases were being randomly selected for human monitoring, then it might be useful if these were the cases that the prudence technique identified as more likely to be unusual. To our knowledge such approaches have not yet been used in industrial systems.

9.5 RDR-BASED MACHINE LEARNING METHODS

In the next section, we will describe developments on combining machine learning and RDR, but before we consider this, we should note that there are in fact a number of machine learning methods based on an RDR representation.

Because RDR builds rules for a case at a time, i.e. for a class at a time, it suggests an approach to machine learning whereby the learner attempts to learn a class at a time. A rule is learned to distinguish cases of a particular class from all other classes. If cases of more than one class fall under the rule, the process is recursively applied to those cases and also to the cases that did not fall under the rule. This is exactly the binary tree structure of SCRDR. A machine learning version of RDR was first developed by Brian Gaines (Gaines 1991a, Gaines and Compton 1992) based on his Induct learning algorithm (Gaines 1989). The RIpple-DOwn Rule learner (RIDOR)[1] is a very widely used implementation of Gaines' Induct-RDR algorithm and is available as part of the WEKA machine learning workbench (Frank et al. 2005, Hall et al. 2009). In Gaines approach, the binomial theorem is used to select rule conditions which give the least likely coverage of cases by that rule. There are a number of other machine learning algorithms that also use the SCRDR representation (Siromoney and Siromoney 1993, Kivinen, Mannila, and Ukkonen 1994, Scheffer 1996). The SCRDR approach developed by Motoda's group will be discussed separately.

Across the standard datasets, on average machine learning with an RDR representation does not seem to have any particular advantage; however, it may have an advantage with an unbalanced dataset. This was suggested in a study comparing learning methods for configuring ion chromatography (Mulholland et al. 1995). A decision tree learner like C4.5 or J48 considers one attribute at a time to decide which attribute-value pair best separates out the classes overall using some sort of information gain or equivalent and does this recursively. As we have commented this may result in classes with small numbers of examples being spread over a number of leaves of the tree which are then discarded at the pruning stage. Induct-RDR in contrast considers one class at a time and tries to find the rule that best separates this class from all other classes using a probability calculation, and does this recursively. This approach may still miss rare classes, but based on her experience, Mulholland suggested that selecting one class at a time, rather than selecting one attribute to separate all classes may be useful for some datasets. Catlett also suggests the interesting possibility that the SCRDR representation may be a more compact mediating representation and in his study transformed other representations into SCRDR (Catlett 1992). Mulholland's study also noted that the Induct-RDR representation was more compact than a C4.5 tree.

9.6 MACHINE LEARNING COMBINED WITH RDR KNOWLEDGE ACQUISITION

Motoda's group has developed methods where both RDR machine learning and human input can be used. When a case is misclassified by an SCRDR tree they use a minimum description length (MDL) calculation to decide the best set of features in the case which differ from the cornerstone case based on having the smallest description length overall. Since a greedy search is used in finding the features, better results may be produced by starting the search with the features the user selects (Wada et al. 2000); that is, the MDL method may find a better set of features starting from the features selected by the user. They also address the problem discussed above where over time a changing population distribution may require a change in the tree and also provide algorithms to delete nodes and prune a tree (Yoshida et al. 2004). They have also used the MDL approach to identify the best default conclusion and also the effect of noise on an MDL-based RDR (Wada et al. 2001). In this work, although a user may be involved, the focus is more on machine learning and the role of the user is simply to select features to use as a starting point for the MDL learner to search for the best set of conditions for the rule. However, there is clearly potential in this approach to further develop integration of machine-learned and expert rules.

In a study by Kim et al. machine learning was applied to very large dataset of factory alarms (Kim et al. 2018). They compared a range of machine learning algorithms including their implementation of a modified version of Induct-RDR, as RIDOR the WEKA version of Induct-RDR was not able to deal efficiently with the very large dataset. The advantage of using their version of Induct-RDR was that it was trivial for rules written by users to override either the overgeneralisation or the overspecialisation of the learned rules and achieve better performance.

Although not using RDR, an email classifier was developed combining both manual rules and machine learning (McCreath, Kay, and Crawford 2006). The best results were obtained when conclusions from hand-crafted rules were chosen over machine learning conclusions if they disagreed. The machine learner tended to produce greater coverage of emails while the hand-crafted rules were more precise.

We have already referred to a semi-supervised learning approach combining RDR and machine learning (Finlayson and Compton 2014). In semi-supervised learning only some cases are labelled. When the learner produced a label that disagreed with an evolving RDR system, the "expert" (a simulated expert) not only labelled the case with the correct label but also added a rule. This extra knowledge resulted in apparently superior semi-supervised learning.

Summarising, the simplest way to combine RDR with a machine learner or anything else, is as wrapper. Data is processed by the machine-learned knowledge base and then processed by an RDR. The default conclusion is the conclusion produced from the machine learner, but any errors in this are corrected by RDR, with rules added as required. This is essentially the approach used by Kim et al. (2018) and McCreath, Kay, and Crawford (2006) although the latter were not using RDR. Earlier in this chapter it was also suggested that this is probably the best way to deal with unbalanced data (that is already labelled). Simply apply whatever machine learner is appropriate, but then correct the errors that will occur for minority classes with RDR rules.

This idea of using an RDR system as a wrapper applies not only to machine learning but can be also applied to any system that processes data in some way to reach conclusions about the data. Appendix 2 includes examples of RDR wrappers to specialize information extraction tools to particular domains and to reduce false positives in a system that identifies duplicate invoices.

Having looked at how RDR can assist machine learning, we will now briefly consider how machine learning can assist RDR.

9.7 MACHINE LEARNING SUPPORTING RDR

Clancey's classic paper on Heuristic Classification (Clancey 1985) argues that most expert systems used an approach of first identifying abstractions of the data and then using combinations of these abstractions to determine the final conclusion. These abstractions could be simple abstractions such as classifying a numerical value as "high", "normal" or "low", but Clancey emphasised that these abstractions could themselves be conclusions of rules, and we have demonstrated this in the GRDR example in Chapter 8. However, with RDR's case by case approach, possible abstractions which would facilitate later knowledge acquisition may be overlooked. Suryanto developed a method to detect possible intermediate abstractions using Muggleton's DUCE algorithm (Muggleton 1992, Suryanto and Compton 2002, 2003, 2004). The conclusion from this work was that possible abstractions could be discovered and the RDR knowledge acquisition effort was significantly reduced when tested on synthetic domains; however, when tested on real domains, few intermediates were found. Suryanto also converted a large real-world PKS MCRDR knowledge base of over 3,000 rules to flat rules, to discover any repetition and possible compression (Suryanto, Richards, and Compton 1999). Only about 10% compression was achieved. This suggests that the need for heuristic classification using rules (rather than function-based abstractions) might not be as great as hypothesised, and perhaps this is related to using RDR. Nevertheless, this work is worth pursuing in real applications as Suryanto's studies showed that automatic discovery of possible abstractions, even if the user rejected all of them, did not add to the knowledge acquisition task for the user.

A second possibility is to use learning to suggest rule conditions. For a chemical pathology report the expert creating a rule has already identified the important features in the case in deciding the appropriate conclusion for the case. Although there might be a fairly large number of attributes, there are likely to be only a few attributes that are relevant to the conclusion. It is a completely different situation for a document classifier, say an email classifier. The user may classify an email simply by sender or topic, but they may also decide on the classification after reading or at least skimming the whole document. What words should be used as rule conditions?

Wobcke's group used a naïve Bayes algorithm to suggest keywords for the user's RDR rules (Ho, Wobcke, and Compton 2003, Wobcke, Krzywicki, and Chan 2008). The approach was evaluated on a large corpus of government emails and outperformed machine learning methods. The words identified by Bayes were highlighted on the email, but it was up to the user, whether they selected any of these or other words. It is also worth noting again that in the evaluation 372 rules were created at an average of

25 secs per rule. The authors also note the disadvantage of a machine learning approach versus an RDR approach in a domain like email where the pattern of emails being received can change over time.

9.8 SUMMARY

The title of this book, "Ripple-Down Rules: The Alternative to Machine Learning" was deliberately provocative given the huge current interest in machine learning. We started with a brief review of the central problem for machine learning of obtaining sufficient high-quality data to learn the concepts required. Those who are experienced with machine learning are well aware of the problem, but developers and companies new to machine learning might be more aware of some of the amazing machine learning algorithms out there and the magic results they have produced, rather than the difficulties of coming up with the data able to do something similar for their particular application.

We then considered the difficulties in acquiring knowledge from domain experts and how a different philosophical analysis leads to an approach like RDR. RDR was presented in considerable detail with worked examples, to try to clear away any assumptions a reader might have about how RDR should work rather than how it actually works. This was followed with a brief discussion of implementation and deployment issues. The central of these is how you validate that a system produces the desired conclusions and it was argued that RDR provides simpler and more realistic validation,

Finally, we returned to machine learning, to examine the problems with suitable datasets in more detail. This led to a brief introduction to machine learning methods using an RDR representation but more importantly to a discussion of how RDR can be used with machine learning. The key suggestion was to use RDR as a wrapper around a machine-learning-based system or in fact any other system that processes data in some way. RDR can be used to correct any errors and improve the performance of the underlying system, perhaps dramatically.

But as a final suggestion on the use of RDR, rather than first trying to assemble the best possible dataset of sufficient size and distribution and with accurate labels for machine learning, if you have someone who already knows how to make the judgements about data, you can use them to very rapidly build an RDR system. The data doesn't have to be clean or well distributed, and it doesn't have to have labels for a domain expert to build rules. The resultant RDR system might be all that is needed, but at very least the feasibility of building a system has been explored as well as data issues – and a consistent data labelling system has been produced in the process. We should finally note again that we are assuming that the domain expert can refer to features in the data that can be readily converted to computer code; in some domains such as image processing this can be so difficult that it can be easier to try to get sufficient accurately labeled data to apply methods like deep learning.

The two appendices are summaries of all the different RDR applications that we know of. The first Appendix describes RDR systems that have been developed in industry and the second, research systems which the authors believe have had substantial validation albeit in a research lab.

NOTE

1 http://weka.sourceforge.net/doc.packages/ridor/weka/classifiers/rules/Ridor.html

Appendix 1

Industrial Applications of RDR

This appendix provides brief summaries of the various industry applications of Ripple-Down Rules (RDR) that we are aware of. A range of other applications based on RDR has been developed, but only reported in research papers. These are presented in Appendix 2.

The information available about these industrial applications of RDR varies greatly. Some companies do not provide information about the technology they use and often the information has come from personal communication.

It is interesting to note that in all the industrial applications we know of, except for YAWL, there was a key person promoting RDR who had had previous direct experience of RDR. The reason this book has been written is precisely to give potential users of RDR some direct experience of RDR.

A1.1 PEIRS (1991–1995)

PEIRS (Pathology Expert Interpretative Reporting System) was in routine use in the Department of Chemical Pathology, St. Vincent's Hospital Sydney for about 3–4 years. It was used to provide interpretative comments for chemical pathology laboratory results. Before PEIRS the only automated comments possible at St. Vincent's were more generic comments produced by simple triggers such as an analyte result being out of the normal range. For more specific comments a pathologist would have to select from a comment bank or write a comment. With PEIRS much more specific comments were produced automatically, but all were checked by a pathologist (Edwards et al. 1993, Preston, Edwards, and Compton 1994) as this was the first deployment of an RDR system for clinical use.

PEIRS used Single Classification RDR (SCRDR) and ended up with about 2,000 rules built by a chemical pathology registrar. About 200 rules were developed offline before the system went into routine use. A report could contain results on any of up to 200 analytes, but normally covered no more than 20 analytes. Reports contained five columns of data representing the last five sets of results for a patient. There were a number of simple operators such as minimum, maximum, average, current and net change that could be used in rule conditions. The laboratory issued about 500 reports a day and comments were provided for about 100 of these in selected subdomains.

PEIRS went out of use when a new hospital information system was deployed which was not capable of referring cases to a system like PEIRS and receiving back clinical interpretations. Although PEIRS was in routine clinical use, it was not commercialised.

Origins: Research on RDR first started at Garvan Institute of Medical Research at St. Vincent's Hospital to deal with the maintenance problems of GARVAN-ES1 (Horn et al. 1985, Compton et al. 1988). PEIRS was initiated by Garvan Institute researchers who had moved to Chemical Pathology at St. Vincent's and also the University of New South Wales.

A1.2 PACIFIC KNOWLEDGE SYSTEMS

Pacific Knowledge Systems (PKS) (http://pks.com.au) was set up following the success of PEIRS. PKS started out developing a generic RDR tool, but with an emphasis on interpreting laboratory data. Its business focus is now almost entirely on interpreting laboratory data and auditing data entry to ensure requests are appropriate. Documents and case studies on the PKS website indicate:

- PKS technology is used at over 90 sites world-wide.
- Over 800 user-developed knowledge bases are in use.
- Over 28 million patient reports were generated annually.
- Across PKS customers over 80% of patient-specific reports are issued without needing human validation.
- In one large laboratory request errors have reduced by 73%.
- RDR-based comments improve detection of familial hypercholesterolaemia (Bender et al. 2016).

- Real-time RDR advice for patients at cardiac risk reduced bed days, readmission, testing and saved money.
- PKS has partnerships with major laboratory technology companies Abbott, Philips and Thermo Fisher.

These knowledge bases are generally developed by chemical pathologists themselves after a few days training from PKS staff on how to use the technology. In the training period PKS staff may develop a "seed" knowledge base, to which the pathologist can continue to add rules. As of 2011 the largest knowledge base had over 12,000 rules (Compton et al. 2011). Compton (2013) shows the evolution of a 3,000 rule knowledge over 8 years. In the first year about 28 hours were spent adding rules, with about 10 hours or less in later years. The median time for a chemical pathologist to add a rule is about a minute or two (Compton et al. 2011). The reason the time to add a rule is so small is because when pathologists want to add a rule they have already identified some feature(s) in the data as the reason to add a rule, so their only task is selecting the feature(s) using the interface provided and if requested also selecting some other discriminating feature(s) to exclude cornerstone cases (Compton et al. 2006, Compton et al. 2011, Compton 2013). Despite the many users building and maintaining knowledge bases, it is always possible that a laboratory can let maintenance lapse, and despite rule building being very simple, skills in this can also lapse.

PKS RDR known as RippleDown is based on MCRDR. It provides a language to enable users to describe and extract features to be then used in rule conditions. It also allows users to set different validation levels for different comments (http://pks.com.au/technology/resources/) as discussed in Chapter 8. A major issue in any PKS deployment is interfacing between PKS technology and the laboratory information system.

Origins: PKS was founded by members of the group who had initiated and worked on PEIRS and other associates[1]. It was initially developed through founder investment, then venture capital funding. It became majority-owned by a private investment company in 2011 and in 2019 was floated on the Australian Stock Exchange.

A1.3 IVIS

Ivis (https://www.ivisgroup.com) provides technology for multi (now omni)-channel retailing. Tesco, famous for its use of multi-channel retailing is one of Ivis's customers. Ivis Sonneto technology is based on an integration of RDR and Conceptual Graphs (Sarraf and Ellis 2006, Ellis 2005, http://jtonedm.com/2008/12/29/first-look-sonetto-retail/). There is no mention of RDR on the current website, but the references above make clear the basis of the technology.

Basically, the rules were used to customise the user experience, e.g. in suggesting products that may be of interest. E.g. if a British user was looking at a holiday in Spain, they may well be interested in a holiday on a Pacific island that was a special offer. A new recommendation like this cannot be made by techniques such as collaborative filtering which depend on examples, but it can be easily made by a business analyst who knows that a British person wanting a holiday in Spain probably wants a holiday in the sun. By 2006 apparently 100s of 1000s of rules had been written (Gerard Ellis, personal communication).

Origins: Ivis developed their RDR system after Gerard Ellis, formerly a PKS software engineer, took up a senior position at Ivis.

A1.4 ERUDINE PTY LTD

Erudine no longer exists, although apparently at one stage it had over 70 staff, but is included here because of their particular application of RDR. It is claimed that Erudine failed not because of the technology, but because of difficulties in getting large debtors to pay their bills (https://www.bloorresearch.com/2014/01/simulating-behaviour-to-replace-legacy/). The technology reappeared in another company, Erudine Financial, but this has also disappeared, although may have morphed into yet another company. There was no acknowledgement that Erudine's Behaviour Engine was based on RDR, despite describing an identical approach; however, the company was previously known RippleDown Solutions and can be found on the Wayback Machine (www.rippledownsolutions.com) and there is a LinkedIn page (https://www.linkedin.com/in/martin-rice-834ba7/) indicating that the CEO of Erudine was also the owner of RippleDown Solutions.

Erudine's main focus was re-engineering legacy software systems. As a legacy system processed cases, each case together with the legacy system outcome for that case would be passed to a business analyst who would write rules to reproduce the outcome for the case using an RDR approach. Although Erudine apparently failed because of debtors, perhaps a challenge for the technology was that some systems to be re-engineered required very complex outcomes, and constructing these was a major task for the business analyst beyond the simple task of identifying differentiating features to build rules. But this is speculation.

Origins: Erudine was founded by a software engineer who had worked for Ivis.

A1.5 RIPPLE-DOWN RULES AT IBM

As discussed earlier, IBM Research carried out a research project on cleansing Indian street address data using a commercial tool, a decision tree and a conditional random fields method and RDR (Dani et al. 2010). The methods all worked similarly on test data from the same area of India as the training data, but the RDR-based method worked much better when applied to all-India data, presumably because the RDR rules reflected an understanding of what an address should be, whereas the learning methods were statistical.

The RDR method used a different single classification RDR for each field in an address. Initial rules were also added independent of cases. Each field also had a dictionary and the rules for that field refer to its dictionary and there can be cases which are fixed with a dictionary update rather than a rule. We point out these adaptations of RDR to highlight that RDR is more a strategy and approach than a particular technology which then needs to be adapted to the target application.

In 2010 this research project was recognised as an IBM "Research Accomplishment". For a research work to be recognized as a Research Accomplishment it must drive new business worth at least USD $10M million and the three team members were recognized with Outstanding Technical Achievement Awards[2]. Apparently, services based on this technology have been sold to numerous customers.

Origins: Ashwin Srinivasan, who as a post-doctoral fellow implemented the original PEIRS system, introduced RDR to IBM Research many years later.

A1.6 YAWL

YAWL (http://www.yawlfoundation.org) is an open source workflow language. It allows for complex data transformations, and integration with other organisational resources and Web Services. It uses RDR to be able to make specific decisions about different parts of the workflow. Single classification RDR is used and there can be many RDR associated with different decision points in the workflow (Adams et al. 2006). There have been more recent extensions to how RDR is used in YAWL (Adams 2013).

Although YAWL is open source it has had significant industry uptake internationally. (http://www.yawlfoundation.org/pages/impact/uptake.html)

Origins: Unlike any other industry application of RDR, the idea of using RDR for YAWL worklets came from a literature search rather than any contact with another RDR project[3].

A1.7 MEDSCOPE

Medscope (https://www.medscope.com.au) provides advice to pharmacists about drug interactions using RDR. Since commercialisation in 2009, over 1200 individual pharmacists have used Medscope's Medication Reviews Mentor and the system has made 800,000 recommendations[4]. Pharmacists use this system voluntarily, so this repeat use clearly suggests the value of its advice. The system also finds more potential drug problems than pharmacists (Curtain et al. 2013).

Origins: Medscope's use of RDR resulted from Ivan Bindoff's PhD project at the University of Tasmania supervised by Byeong Kang.

A1.8 SEEGENE

SeeGene (https://www.seegenemedical.com) is a very large laboratory that recently started providing RDR-based diagnostic reports for about 200 small hospitals as well as GPs in Korea. Seegene also founded Sciosoft Technology to focus on RDR applications. Seegene uses an RDR engine developed by Sciosoft for protein electrophoresis and peripheral blood smear tests. They propose to expand their use of RDR to other reporting areas.

Since this is being written in 2020, it is worth noting Seegene is well known as having developed a real-time test for SARS-CoV-2, the Allplex 2019-nCoV assay, used in over 60 countries.

Origins: Seegene's use of RDR resulted from a collaborative project again with Byeong Kang at the University of Tasmania and the Korean Advanced Institute of Science and Technology, funded by a grant from the Korean Ministry or Trade, Industry and Energy.

A1.9 IPMS

IPMS (http://www.stable.co.kr), which started in 2010, provides RDR-based diagnostic advice about system alerts and at the time of writing has 100 customers.

Origins: IPMS's use of RDR resulted from a collaborative project with Byeong Kang at the University of Tasmania.

A1.10 TAPACROSS

Tapacross (https://www.tapacross.co.kr) provides social media trend prediction services. RDR is used to classify very large number of documents.

Origins: Tapacross's use of RDR resulted from a collaborative project with Byeong Kang at the University of Tasmania.

A1.11 OTHER

In Korea, RDR has also been used in some of the internal processes of a very large steel company, for failure reporting, analysis and corrective action, but this is under confidentiality agreements. Also, an RDR engine by BiZin Technology (http://bizintech.co.kr/eng) has been applied to a chatbot interface for a car navigation system, with the service released in Korea in 2018, but the information about the car manufacturer is also confidential.

There may also be other industry use of RDR that we are not aware of, because similar to Erudine, the companies may have chosen not to identify that they use RDR.

Clearly RDR has had significant and successful industry uptake. However, it is interesting to note that all of these developments, apart from YAWL, came about by some direct contact with other RDR research, not from reading the literature, although there are many RDR papers. This seems to confirm our own experience that people tend to misunderstand the very simple ideas in RDR and assume it must be doing something different – unless they have had some direct experience of RDR. Hopefully, the worked examples in this book will help in this regard.

NOTES

1 Paul Compton previously had a small shareholding in PKS, but is no longer a shareholder.
2 Personal communication from L. Venkata Subramaniam (IBM)
3 Personal communication from Michael Adams, the YAWL developer
4 Personal communication Medscope

Appendix 2

Research-Demonstrated Applications

This chapter provides summaries of Ripple-Down Rules (RDR) applications that have been demonstrated in research projects but have not been put into routine use outside the laboratory. The purpose of this chapter is to suggest to readers interested in going further with RDR, possible application areas that research has already established. We have selected areas that seem more likely to us to be of practical interest rather than more theoretical research, and apologise to those authors whose papers we have missed. The level of validation in these studies varies from substantial to preliminary results but with practical potential. Richards' review of 20 years of RDR research also covers much of the material here from before 2009 (Richards 2009).

A2.1 RDR WRAPPERS

We have already discussed the idea of an RDR wrapper in Chapter 9. All inputs and outputs from the underlying system are passed to the RDR system. If no rule applies, the default conclusion is the output of the underlying system, but RDR rules can be added to modify the output from the underlying system if and when required. A few RDR wrapper applications have been developed.

A2.1.1 SAP Duplicate Invoices

Duplicate invoices are a major problem for business, with estimates that 2% of all payments are for duplicate invoices (Dubinsky and Warner 2008). SAP had a system for detecting duplicate invoices. Apparently, it successfully detected all duplicates, but the false positive rate was 92%; that is, only 8% of the suspect invoices that account clerks had to check by hand were actually duplicate invoices. An RDR system was developed to reduce the false positives (Ho et al. 2009).

About 4,000 cases in chronological order were inspected by the researcher involved and about 50 rules added. These rules were then tested on about 5,000 unseen cases. The false positive rate dropped from 92% to 19%, obviously a huge improvement. There is an apparent cost of manually going through the 4,000 cases; however, the account clerks have to do this anyway, so adding rules for 50 of these cases would be a trivial extension to the workload – even if the account clerks had to pass the 50 cases to someone more senior to add the rules. And of course, rules could be added for cases beyond the 4,000 looked at in this project. It is not known if SAP ever deployed such a system.

A2.1.2 RDR Wrapper for the Stanford Named Entity Recogniser

Named entity recognisers work well on well-formed text such as journalistic text, but less well on the more informal text found in many web documents. In these experiments the Stanford NER was tested on a Web dataset. The F-score was 76.96%, but with an RDR wrapper and after adding 83 rules over four hours to cover 200 training sentences the F score for 341 other sentences from the same domain was 90.48% (92.12% precision and 88.68% recall) (Kim and Compton 2012). This is about the same as the performance of the Stanford NER on the formal CONLL corpus (F score of 90.8%).

The strength of the Stanford NER is that it is a general tool for any domain. The addition of the RDR wrapper does not improve its general performance, but only its performance in the subdomain for which RDR rules were added. However, in real applications, tools such as the Stanford NER are likely to be used for specific domains of interest, and what these results show is that a few hours work building an RDR wrapper gives a substantial improvement in named entity recognition in the domain of interest. Although not tested in this research, it is assumed that the RDR wrapper, while improving Stanford NER's performance in the domain of interest, probably does not degrade its general performance.

A2.1.3 RDR Wrapper for REVERB

The same approach was used to improve REVERB's extraction of relations from informal Web documents (Kim and Compton 2012). When tested on informal Web text rather than well-formed text, REVERB performed poorly in extracting relations with a recall of 45% and a precision of 41% in identifying relations. 53 RDR rules were added in 2 hours and the resulting performance on test sentences from the same domain was an 81% recall and 90% precision.

These are extraordinarily good results for improving the performance both of the Stanford NER and REVERB on informal web text in a domain of interest. Perhaps the text datasets selected had particular characteristics; perhaps they were very specific in a way that was difficult for the general tools, but easily corrected with relatively few RDR rules. This is possibly the case, but these were not specially constructed datasets, and the research they come from is well known. The training dataset came from Bunescu and Mooney (2007) with 250 citations while the test dataset, extracted by the same mechanism came from Banko and Etzioni (2008) with 415 citations. There were no sentences in common between the test and training data. Even if the particular nature of the datasets facilitated such good results, the results are nevertheless highly suggestive that RDR wrappers are likely to be of value in improving the performance of general tools in specific domains.

A2.1.4 Machine Learning and RDR Wrappers

We have already covered this in some detail in Chapter 9. Applications included a factory fault detection system (Kim et al. 2018) and an email classifier (McCreath, Kay, and Crawford 2006); although this did not use RDR.

A2.2 TEXT-PROCESSING, NATURAL LANGUAGE PROCESSING AND INFORMATION RETRIEVAL

This section also includes question answering, conversational agents and help desks as these are all essentially about retrieving the right information in response to a query.

A2.2.1 Relationship Extraction

Two of the RDR wrapper studies were also examples of RDR-based open information extraction. This work followed earlier work successfully using RDR alone for information extraction in specific web domains (Kim, Compton, and Kim 2011). This work has recently been followed up (Kim 2017).

In contrast to the previous work, the web page dataset for Kim's 2017 study consisted of educational institutions, government organisations and industry and the web pages were selected only on the basis of url (.com, .gov, .edu); there was no key word selection. The study was based on extracting the specific relations: "*hasPhone*", "*hasFax*", "*hasAddress*", "*hasEmail*" and "*hasDomainAddress*". These were then further categorised to distinguish organisations and persons, e.g. "*O_hasPhone*" and "*P_hasPhone*". 100 pages were randomly selected as training data and another 100 as test data. 22 rules were added over 7 hours, which also involved inspecting 5770 tuples. There was no extra preparation time labelling data or understanding the data structure in advance. The 100 test pages were then looked at in detail. The recall was 83% and the precision 93%.

In contrast to the previous studies, these results were purely from RDR, no general tools were involved. Also, there was no key word selection of web pages, but specific relations were considered. The RDR rGALA tool was built on GALA, an in-house tool which provided part-of-speech tagging and named entity recognition, developed by the Australian Defence Science and Technology Group (DST). There was of course a cost in developing rGALA as with any other tool, but for the actual experiment there were no other costs involved in data preparation or understanding apart from the 7 hours knowledge acquisition. Clearly this is a cost-effective way of building systems to extract relations of interest.

A2.2.2 Temporal Relations

Pham and Hoffman developed the KAFTIE system based on RDR (Pham and Hoffmann 2006). In experiments on temporal data from the Timebank corpus this approach outperformed machine learning methods, as used by these researchers, with as little as 40 rules. Before getting to the RDR stage KAFTIE also used standard tools from GATE (Cunningham et al. 2002) for basic language processing tasks.

A2.2.3 Positive Attributions

Pham and Hoffman previously used the KAFTIE system for extracting positive attributions from scientific papers (Pham and Hoffmann 2004). Two days of the user adding rules produced a precision >74% and a recall >88% in a prototype system to identify positive attributions in scientific papers. A precision of > 74% considerably outperformed baseline machine learning algorithms trained on the same data.

A2.2.4 Part of Speech Tagging

Xu and Hoffmann developed an RDR/machine learner for text processing (Xu and Hoffmann 2010). The output from Brill's transformational learning was transformed to an RDR and further refinement rules added. The paper notes that 60 hours of knowledge acquisition resulted in a POS tagger that slightly exceeded state of the art taggers with over 15 years of development. This work could also be characterised as an RDR wrapper approach except that the Brill learner output was transformed into an RDR tree.

Nguyen et al. (Nguyen et al. 2011) also developed an RDR approach based on Brill's method. The key difference from Xu and Hoffmann's approach is that the RDR rules are selected automatically, so the approach exploits the refinement structure of single classification RDR rather than human knowledge acquisition. The approach produced the best results (by a small margin) on the Penn Treebank corpus and also the Vietnamese treebank corpus. The method has now been applied to a wide range of languages, with comparable performance to the current best methods, but with very fast training and tagging times (Nguyen et al. 2016).

This approach of using the RDR refinement structure, but without human knowledge acquisition, has also been applied to Vietnamese word segmentation (Nguyen et al. 2017a).

A2.2.5 Named Entity Recognition

RDR has also been applied to Vietnamese named entity recognition (Nguyen and Pham 2012). This paper is particularly interesting in that it makes a direct comparison with a conventional rule approach applied to the same corpus. As would be expected the conventional method gets more difficult as more rules are added.

A2.2.6 Extracting Processes

Process descriptions, even as simple as recipes, describe the same process in many different ways. In this project RDR was used with standard NLP tools to extract consistent process descriptions (Zhou et al. 2016).

A2.2.7 Email Classification

We have already discussed this work in Chapter 9. Wobcke et al. developed a system for classifying email into folders using RDR and a Bayesian prediction of which words might be more useful as rule conditions and which folders were more likely, but the user was free to choose any words or folders (Ho, Wobcke, and Compton 2003). An evaluation study with an industry partner meant that almost 17,000 pre-classified emails were available for evaluation (Wobcke, Krzywicki, and Chan 2008). It was known that there were errors in the classifications, so a target of 75% accuracy was set covering 50 classifications. There was a further constraint in that no more than 10 minutes would be spent constructing rules for each classification; that is, the evaluation could take no more than 500 minutes and assuming 30 secs per rule, there would possibly be about 1,000 rules. In fact, an accuracy of 85% was achieved by adding 372 rules over 8–9 hours for an initial training set of 7,000 emails and the average time per rule was 25 secs. Various machine learning methods were also evaluated (Krzywicki and Wobcke 2009). The manual RDR produced much better rules, particular in the early stages with fewer training data.

Cho and Richards also developed an RDR-based email classifier (Cho and Richards 2004). However, in their approach a user did not select rule conditions, rather rules were constructed automatically using keywords suggested by a Bayes threshold method as rule conditions. This method outperformed the other automated methods they evaluated.

A2.2.8 Web Monitoring

The web is often monitored for topics of interest, but standard keyword monitoring retrieves too many items. In this research standard keyword monitoring was followed by MCRDR classification to provide a much finer classification of the retrieved pages. For example, if "diabetes" was being monitored then the pages retrieved could be classified as "glucose monitoring", "pumps", "retinopathy" etc. Research covered both push and pull monitoring, and also using MCRDR knowledge to help guide search. It is not possible to provide quantitative comparisons with other methods, but clearly RDR provided for more precise, but still very easy web monitoring (Park, Kim, and Kang 2003, Kim, Park, Deards, et al. 2004a, Kim, Park, Kang, et al. 2004b, Kim et al. 2009, Kang and Kim 2009, Kang et al. 2006, Kang 2010).

A2.2.9 Other Language Tasks

To facilitate string similarity measurement RDR has been used for processing graphemes and phonemes for the Burmese language (Wai et al.).

They have also been used to produce consistent Latin script versions of Burmese proper names (Zaw et al. 2020).

A2.3 CONVERSATIONAL AGENTS AND HELP DESKS

A2.3.1 Conversational Agents

The same group who worked on Vietnamese text processing above have also developed an ontology-based question answering system where RDR rules are used to recognise the type of question and convert it to an appropriate intermediate representation to provide an answer (Nguyen, Nguyen, and Pham 2017a). The paper showed high accuracy, 82.4% for Vietnamese questions. Because rules can be written for any language they also developed an English system, and concluded from 12 hours of adding rules that the approach allowed for very rapid development of useable systems.

Other RDR conversational research developed a version of MCRDR, C-MCRDR, which maintains the context of a conversation (Herbert and Kang 2018). A last-in-first-out stack of frames is maintained where each frame contains a set of satisfied rules from a previous conversational request. Each frame contains a parent rule and perhaps refinement rules (standard MCRDR). Assuming the parent rule is satisfied, each child rule is evaluated and if none is satisfied the next frame is considered and so on until no frames are available and the last conclusion is used. An evaluation study was based on answering question about 34 distinct ICT courses, with 140 individual assessments and 23 teaching teams, which provided a good test for maintaining context. The scope of the question answering was shallow but the results were excellent in that although only 80.3% of questions were answered satisfactorily, only 4.4% of the errors were due to incorrect dialogue classifications, the rest being due to out of scope questions. The most striking result was a 97% reduction in the rule count compared to an estimated count of the rules needed without maintaining context. The evaluation also included speech recognition.

The same system is the basis of the BiZin Technology chatbot for car navigation referred to in Appendix 1. The stack approach enables the system not only to understand and respond to requests in context but also to seamlessly change context, e.g. from navigation directions, to music selection to climate control in the car.

Current work on RDR conversational agents is focused on improving text to speech transcription given a rules-based context (Herbert 2020 – to be submitted).

A2.3.2 Help Desks

The initial work on using RDR for help desks in essence had three methods for retrieving documents (Kang et al. 1997). The first was standard information retrieval: entering keywords. The second used an MCRDR tree to ask the user about specific keywords – which were conditions in rules. The third was a hybrid which allowed the user to freely enter keywords and then the MCRDR tree was used to ask the user for keywords which distinguished the documents. The user was asked whether the appropriate information was retrieved and if not, all the information about the interaction was saved and passed to a domain expert to add more rules. This approach was not tested with real users.

Vazey and Richards researched a much more challenging help desk problem – a multi-national company call centre providing ICT support (Vazey and Richards 2004, 2005). With this type of service, to resolve a problem might require multiple calls perhaps over weeks with details such as: who raised the call, the product involved, environmental variables and the current status of the call. Typically, such help desk systems track this information, but do not capture the problem-solving knowledge of the expert nor their expertise in searching for a solution.

For this problem a case continues over time with more data added until it is resolved. Similar to the Kang et al. help desk above, if a rule is reached with missing conditions, these may be requested. The system also provided for conclusions to be used as conditions in other rules, and conclusions could be of different types. The system also provided for backtracking to establish if a rule could fire. These are just a few of the elements of a quite complex system.

The approach underwent a trial with real users (Vazey and Richards 2006b). After 7 hours effort the team had dealt with 105 cases and 55 rules, which solved 90% of equipment errors in the chosen subdomain, which were 30% of the errors seen by the system, and these errors were 20% of the ~ 5,000 cases seen by the global ICT support centre. In summary after less than an hour's training followed by 7 hours knowledge acquisition effort users constructed a knowledge base able to provide solutions automatically for more than 270 cases per day without requiring trouble-shooters to work out the type of problem or to search for a solution in the corporate data base. The trial ceased after 107 rules and 172 cases. Although this was a significant and successful trial in a real-world application, it is not known how far the system was eventually developed or whether it went into routine use.

A2.4 RDR FOR OPERATOR AND PARAMETER SELECTION

There are many applications where at various stages of a process the user has to choose which operator to apply or which value to use in an operator in part of the process. Such selection is normally a combination of trial and error and experience developed through previous trial and error. The idea in this research was to use RDR to learn why the user makes a particular selection of operator or value. The general conclusion from these studies is not that you get a better final outcome for a given problem, rather you get a solution as good as the best published, but far more quickly and easily.

A2.4.1 Genetic Algorithms

Developing a genetic algorithm involves creating then trialing mutation and fitness functions that select which mutations will survive. Bekmann and Hoffman (Bekmann and Hoffmann 2004) proposed using a general-purpose genetic algorithm but then rather than waiting until the algorithm terminated, they proposed to stop it periodically and inspect the mutations and their survival and where appropriate write rules to change the mutation and fitness functions. The initial paper found that this approach enabled them to solve state-of-the-art VLSI routing problems with a general-purpose algorithm. The approach was then proposed as a general solution for heuristic search including extra features such as a trigger function to determine when to stop the process for inspection (Bekmann and Hoffmann 2005). It took about 30 hours to add rules to convert a generic genetic algorithm into a solution that performed comparably to industry standard algorithms that took years to develop. The NRDR extension of RDR discussed in Chapter 7 was used in this project.

A2.4.2 Image Processing

Image processing normally involves a range of steps, thresholding, thinning, edge detection, techniques to assess texture etc. Typically, various parameters will be trialed, but then if later steps in the overall process do not produce satisfactory results, parameters may be changed for any of the earlier steps and the process repeated until finally an adequate solution is produced. Instead of simply iterating through a series of parameter changes Misra wrote rules as to why parameters were changed (Misra, Sowmya, and Compton 2006). In their lung edge detection project they developed a system which “quickly learned to match and exceed the performance of a fixed algorithm system handcrafted by a vision expert". This idea was generalised to more complex image processing problems in which there was a large network of processes involved, and more detailed evaluation data was presented (Misra, Sowmya, and Compton 2010). Given the success of deep learning in image processing it would be interesting to see if this approach could also be used in helping develop a deep learner for a particular image application.

A2.4.3 Configuring and Selecting Web Resources

Given the huge array of web services becoming available, this research was essentially about selecting appropriate combinations of services. One area of this research has concerned selecting the appropriate similarity detection tools for a given task (Ryu and Benatallah 2012, 2017, Ryu et al. 2011). The conclusion from this work is that using RDR to select similarity detection tools gives as good similarity results on standard datasets as the best published methods, but with less effort and fewer rules.

Barukh and Benatallah (2014) developed a web service process management tool using RDR. In the evaluation, this tool outperformed industry standard tools in terms of usability, productivity and performance.

A2.4.4 Autopilots

Shiraz and Sammut (Shiraz and Sammut 1997) developed an autopilot (for a flight simulator) based on allowing pilots to write RDR rules on what to do in flying a plane. However, the fine control in what a pilot does cannot be verbalised; e.g. a pilot landing a plane is constantly making very small corrections. Machine learning is used to learn these fine motor skills. In the wrapper section above RDR was used to correct machine learning for particular situations. Here the reverse happens with machine learning being used to refine manually added rules.

A2.4.5 Multiple Agent Games

A system was developed to play robot soccer in the simulation league (Finlayson and Compton 2010). Although the system did not compete in any actual competitions, four coaches of different abilities and backgrounds trained teams able to play at the same level as the finalists in the Robocup 2007 2D simulation tournament. An earlier version of the GRDR approach in Chapter 7 was used which allowed for repeat inference using previous conclusions as rule conditions. A single classification SCRDR tree was used, but each time a conclusion was reached it was added to working memory and the tree reprocessed. The system was multi-agent in that each player had its own world model and ran the knowledge base against that world model. The major difficulty in this work was providing a comprehensive and useable language for the coaches to build rules and the resulting rule language consisted of 194 attributes and functions. This included concepts such as the closest teammate, the number of opponents in a given region of the field and the distance from the player to particular points on the field.

A2.4.6 Configuration

A prototype system was developed to configure ion-chromatography analyses (Ramadan, Mulholland, et al. 1998a, Compton et al. 1998). This is essentially a configuration task with multiple interdependent parameter choices: gel type, solute, detector, column etc. In all there are about 10^{12} possible selections of parameters and values. In contrast to the soccer agents who used repeat inference SCRDR, repeat inference MCRDR was used. As this may result in multiple differing conclusions for the same attribute, the simple strategy was used of only adding a conclusion to working memory for the next inference if only one conclusion was given (perhaps multiple times) for that attribute. On a fairly small evaluation of 81 cases this approach provided a workable ion chromatography solution for 71 cases (88%).

A2.4.7 Energy Efficiency

RDR has been used to capture how experienced train drivers drive trains in a more efficient way than inexperienced drivers. The rules were based on studying records of experienced drivers, identifying patterns and encoding these as RDR; a fuzzy strategy tree was then used to control the train smoothly. It is claimed the research was evaluated on commercial trains resulting in an energy saving of over 10% (Huang et al. 2019).

A2.5 ANOMALY AND EVENT DETECTION

A2.5.1 Network Intrusion and Bank Fraud

The approach in this project (Prayote and Compton 2006, Prayote 2007) was to use a fairly simple probabilistic function to detect if a value was outside a range, given the number of values already seen that fell inside the range. Any value that was not identified as an outlier was added to the values used for the probabilistic calculation so that the range became more tightly defined as time went on. However, one cannot use a single function for network traffic as it changes during the day, at special times of the year, with certain events etc. RDR was used to specify the different regions of network traffic, and this could be constantly refined as traffic that was identified as anomalous, but was actually legitimate, was covered by a new function for the defined region. The approach performed comparably to other network intrusion detection methods on real network data. It was also evaluated on a non-network domain and drastically reduced false positives (unnecessary alerts) while missing very few anomalous cases.

A similar approach but combined with another method for identifying anomalies was used in detecting internet bank fraud (Maruatona, Vamplew, and Dazeley 2012b, Maruatona 2013) and produced results comparable to commercial systems. The idea of prudent knowledge bases that recognise when something is anomalous was discussed in Chapter 9.

A2.5.2 Stock Market Events

In contrast to the studies above aimed at finding outliers that indicated some sort of attack or fraud, Chen and Rabhi looked at cleaning up financial time series data to provide more useful data (Chen and Rahbi 2013, Chen and Rahbi 2016). For example, if two dividend payments had the same date one might be a duplicate.

A2.6 RDR FOR IMAGE AND VIDEO PROCESSING

RDR cannot provide the type of image processing deep learning excels at; however, it may provide a useful adjunct.

A2.6.1 RDR and Machine Learning for Image Processing

Simple machine learning can abstract slightly higher-level features such as colour and texture from low-level pixel features. Information such as area, shape, direction of principle axis etc. can be readily extracted from regions of the same feature. Of course, such information will only be approximate, but if rules are then used to consider relationships with other regions, a knowledge base can be readily constructed to identify where a robot is in a given space (Kerr and Compton 2003). An RDR decision-tree structure can also be used to support simple machine learning. When a region is misclassified and should have a different classification the positive training examples for the new class are the pixels from the misclassified region and the negative training example are the pixels used as positive examples in learning the previous class. This approach results in a standard SCRDR exception structure and is a much simpler approach than trying to relearn the class that resulted in the misclassification (Kerr and Compton 2002). The evaluation in this research was limited.

A similar approach was used for a Robot soccer vision system, except that an RDR machine learner was used throughout to learn both low-level features like colour and also more complex features such as one colour on top of another (Pham and Sammut 2005). The method was applied to learning to identify objects such as the goal and beacons (one colour on top of another) marking different corners and the half-way line. The approach was extended further to allow users to verbally identify objects and develop recursive concepts (D'Este and Sammut 2008).

A2.6.2 RDR for Image Processing Parameters

This approach was discussed above in the section on operator selection and adjustment. The normal approach to building an image processing system is to vary parameter values and operators used at different stages of the image processing until a good solution is arrived at. A user will be able to articulate at least partially, the reasons for their choices. If RDR is used to capture these choices, overgeneralised or inappropriate justifications can be rapidly refined (Misra, Sowmya, and Compton 2010).

A2.6.3 RDR for Combining Image Primitives

In this research standard image processing techniques were used to extract primitives such a horizontal stroke, vertical, diagonal etc. from Chinese characters, and the user then wrote rules referring to these primitives to identify the different characters (Amin et al. 1996).

A2.6.4 Video Processing

Similar to some of the image analysis research above, RDR was applied to previously extracted video features (Sridhar, Sowmya, and Compton 2010, Sridhar 2011). The expert's task was simply to identify the features to be extracted, while the values were provided automatically from upper and lower bounds. Features extracted included head position and centre of gravity and voxel models for standing, crouching, jumping, and walking. Cases consisted of a number of sequential frames. Evaluation in this sort of application was complex, but clearly demonstrated the system learned to recognise gestures.

References

Adams, M. 2013. "Usability Extension for the Worklet Service." *YAWL Symposium*. edited by A. Hense, T. Freytag, J. Mendling and A. ter Hofstede, 69–75: CEUR Workshop Proceedings.

Adams, M., A. ter Hofstede, D. Edmund and W. van der Aalst. 2006. "Worklets: a service-oriented implementation of dynamic flexibility in workflows." *Proceedings of the 14th International Conference on Cooperative Information Systems (CoopIS'06)*, Montpellier. edited by R. Meersman and Z. Tari, 291–308: Springer-Verlag.

Amin, A., S. Anwar, B. Shah and A. M. Khattak. 2017. "Compromised user credentials detection using temporal features: a prudent based approach." *Proceedings of the 9th International Conference on Computer and Automation Engineering*, Sydney, 104–110: ACM.

Amin, A., M. Bamford, A. Hoffmann, A. Mahidadia and P. Compton. 1996. "Recognition of Hand-printed Chinese Characters using Ripple-Down Rules." In *Advances in Structural and Syntactical Pattern Recognition*, edited by P. Perner, P. Wang and A. Rosenfeld, 371–380. Berlin: Springer-Verlag.

Amin, A., F. Rahim, M. Ramzan and S. Anwar. 2015. "A prudent based approach for customer churn prediction." *11th International Conference: Beyond Databases, Architectures and Structures, BDAS 2015*, Ustroń. edited by S. Kozielski, D. Mrozek, P. Kasprowski, B. Malysiak-Mrozek and D. Kostrzewa, 320–332: Springer.

Amin, A., B. Shah, S. Anwar, F. Al-Obeidat, A. M. Khattak and A. Adnan. 2018. "A prudent based approach for compromised user credentials detection." *Cluster Computing* 21 (1):423–441.

Ammanath, B., S. Hupfer and D. Jarvis. 2019. "Thriving in the Era of Pervasive AI: State of AI in the Enterprise, 3rd Edition." *Deloitte*

Banko, M. and O. Etzioni. 2008. "The tradeoffs between open and traditional relation extraction." *Proceedings of ACL-08: HLT*, Colombus, 28–36: Association for Computational Linguistics.

Barukh, M. C. and B. Benatallah. 2014. "ProcessBase: a hybrid process management platform." *International Conference on Service-Oriented Computing, ISCOC 2014*, Paris. edited by X. Franch, A. K. Ghose, G. A. Lewis and S. Bhiri, 16–31: Springer.

Bekmann, J. and A. Hoffmann. 2004. "HeurEAKA–A New Approach for Adapting GAs to the Problem Domain." *Pacific Rim International Conference on Artificial Intelligence*, Auckland. edited by C. Zhang, H. W. Guesgan and W. K. Yeap, 361–372: Springer.

Bekmann, J. P. and A. Hoffmann. 2005. "Improved Knowledge Acquisition for High-Performance Heuristic Seach." *IJCAI-05, Proceedings of the 19th International Joint Conference on Artificial Intelligence*, Edinburgh, 41–46: Morgan Kauffmann.

Bender, R., G. Edwards, J. McMahon, A. J. Hooper, G. F. Watts, J. R. Burnett and D. A. Bell. 2016. "Interpretative comments specifically suggesting specialist referral increase the detection of familial hypercholesterolaemia." *J Pathology* 48 (5):463–466.

Bennett, C. 2018. "Watson Recommends Incorrect Cancer Treatments, System Training Questioned." *Clinical OMICs* 5 (5):29–29.

Beydoun, G. and A. Hoffman. 1997. "Acquisition of search knowledge." *Knowledge Acqusition, Modeling and Management, EKAW'97*, Sant Feliu de Guixols. edited by E. Plaza and R. Benjamins, 1–16: Springer.

Beydoun, G. and A. Hoffman. 1998. "Building problem solvers based on search control knowledge." *11th Banff Knowledge Acqusition for Knowledge-Based Systems Workshop*, Banff. edited by B. Gaines and M. Musen, SHARE 3, 1–18: *SRDG Publications, University of Calgary.*

Beydoun, G. and A. Hoffman. 1999. "Hierachical Incremental Knowledge Acquisition." *12th Banff Knowledge Acqusition for Knowledge-Based Systems Workshop*, Banff. edited by B. Gaines, R. Kremer and M. Musen, 7-2.1–7-2.20: SRDG Publications, University of Calgary.

Beydoun, G. and A. Hoffmann. 2000. "Incremental Acquisition of Search Knowledge." *International Journal of Human Computer Studies* 52 (3):493–530.

Beydoun, G. and A. Hoffmann. 2001. "Theoretical basis for hierarchical incremental knowledge acquisition." *International Journal of Human Computer Studies* 54 (3):407–452.

Bindoff, I. and B. Kang. 2011. "Applying Multiple Classification Ripple Round Rules to a Complex Configuration Task." *AI 2011: Advances in Artificial Intelligence*, Perth. edited by D. Wang and M. Reynolds, 481–490: Springer

Bobrow, D., S. Mittal and M. Stefik. 1986. "Expert systems: perils and promise." *Communications of the ACM* 29 (9):880–894.

Boeker, M., I. Tudose, J. Hastings, D. Schober and S. Schulz. 2011. "Unintended consequences of existential quantifications in biomedical ontologies." *BMC Bioinformatics* 12 (1):456.

Buchanan, B. 1986. "Expert systems: working systems and the research literature." *Expert Systems* 3:32–51.

Bunescu, R. and R. Mooney. 2007. "Learning to extract relations from the web using minimal supervision." *Proceedings of the 45th Annual Meeting of the Association of Computational Linguistics*, Prague. edited by A. Zaenen and A. van den Bosch, 576–583: Association of Computational Linguistics.

Burns, E. M., E. Rigby, R. Mamidanna, A. Bottle, P. Aylin, P. Ziprin and O. Faiz. 2011. "Systematic review of discharge coding accuracy." *Journal of Public Health* 34 (1):138–148.

Cabitza, F. and G. Banfi. 2018. "Machine learning in laboratory medicine: waiting for the flood?" *Clinical Chemistry and Laboratory Medicine (CCLM)* 56 (4):516–524.

Cao, T., E. Martin and P. Compton. 2004. "On the Convergence of Incremental Knowledge Base Construction." *Discovery Science, 7th International Conference, DS 2004* Padova. edited by E. Suzuki and S. Arikawa, 207–218: Springer.

Cao, T. M. and P. Compton. 2005. "A Simulation Framework for Knowledge Acquisition Evaluation." *Twenty-Eighth Australasian Computer Science Conference (ACSC2005)*, Newcastle. edited by V. Estivill-Castro, 353–360.

Cao, T. M. and P. Compton. 2006a. "Evaluation of Incremental Knowledge Acquisition with Simulated Experts." *AI 2006: Advances In Artificial Intelligence, 19th Australia Joint Conference on Artifical Intelligence*, Hobart, Australia. edited by A. Sattar and B. Kang, 39–48: Springer.

Cao, T. M. and P. Compton. 2006b. "Knowledge Acquisition Using Simulated Experts." *Managing Knowledge in a World of Networks, 15th International Conference EKAW 2006*, Podebrady, Czech Republic. edited by S. Staab and S. Vojtech, 35–42: Springer.

Catlett, J. 1992. "Ripple-Down-Rules as a Mediating Representation in Interactive Induction." *Proceedings of the Second Japanese Knowledge Acquisition for Knowledge-Based Systems Workshop*, Kobe, Japan. edited by R. Mizoguchi, H. Motoda, J. Boose, B. Gaines and R. Quinlan, 155–170: JSAI.

Chavis, S. 2010. "Coding's Endgame." *For The Record* 22 (17):18.

Chen, W. and F. Rahbi. 2013. "An RDR-based approach for event data analysis." *Service Research and Innovation, Third Australian Symposium (ASSRI 2013)* Sydney. edited by J. Davis, H. Demirkan and H. Motahari-Nezhad, 1–14: Springer.

Chen, W. and F. Rahbi. 2016. "Enabling user-driven rule management in event data analysis." *Information Systems Frontiers* 18 (3):511–528.

Cho, W. and D. Richards. 2004. "E-Mail Document Categorization Using Bayes TH-MCRDR Algorithm: Empirical Analysis and Comparison with Other Document Classification Methods." *Pacific Rim International Conference on Artificial Intelligence (PRICAI 2004)*: Auckland, Berlin. edited by B. Kang, A. Hoffmann, T. Yamaguchi and W. Yeap, 226–240: Springer.

Choi, Y., J. Chung, K. Kim, K. Kwon, Y. Kim, D. Park, S. Ahn, S. Park, D. Shin and Y. Kim. 2019. "Concordance rate between clinicians and Watson for oncology among patients with advanced gastric cancer: early, real-world experience in Korea." *Canadian Journal of Gastroenterology and Hepatology*: Article ID 8072928.

Clancey, W. J. 1985. "Heuristic classification." *Artificial Intelligence* 27:289–350.

Clancey, W. J. 1997. *Situated Cognition: On Human Knowledge and Computer Representations (Learning in Doing-Social, Cognitive and Computational Perspectives)*: Cambridge: Cambridge University Press.

Colomb, R. M. 1999. "Representation of Propositional Expert Systems as Partial Functions." *Artificial Intelligence* 109:187–209.

Compton, P. 2013. "Situated cognition and knowledge acquisition research." *International Journal of Human-Computer Studies* 71:184–190.

Compton, P., T. Cao and J. Kerr. 2004. "Generalising Incremental Knowledge Acquisition." *Proceedings of the Pacific Knowledge Acquisition Workshop 2004*, Auckland. edited by B. Kang, A. Hoffmann, T. Yamaguchi and W. K. Yeap, 44–53: University of Tasmania Eprints repository.

Compton, P., R. Horn, R. Quinlan, L. Lazarus and K. Ho. 1988. "Maintaining an expert system." *Proc 4th Aust Conf on applications of expert systems*. edited by J. R. Quinlan, 110–129: The University of Technology, Sydney.

Compton, P. and R. Jansen. 1988. "Knowledge in context: A strategy for expert system maintenance." *Australian Joint Conference on Artificial Intelligence (AI 88) Lecture Notes in Computer Science vol 406*, Adelaide. edited by C. J. Barter and M. Brooks, 292–306: Springer.

Compton, P. and R. Jansen. 1990. "A Philosophical Basis for Knowledge Acquisition." *Knowledge Acquisition* 2:241–257.

Compton, P., Y. S. Kim and B. H. Kang. 2014. "Linked Production Rules: Controlling Inference with Knowledge." *Knowledge Management and Acquisition for Smart Systems and Services, PKAW 2014*, Gold Coast. edited by Y. S. Kim, B. H. Kang and D. Richards, 84–98: Springer.

Compton, P., L. Peters, G. Edwards and T. G. Lavers. 2006. "Experience with Ripple-Down Rules." *Knowledge-Based System Journal* 19 (5):356–362.

Compton, P., L. Peters, T. Lavers and Y. Kim. 2011. "Experience with long-term knowledge acquisition." *Proceedings of the Sixth International Conference on Knowledge Capture, KCAP 2011*, Banff, 49–56: ACM (a version of this paper with minor corrections is at https://pks.com.au/wp-content/uploads/2015/03/WhitePaperExperiencewithKnowledgeSystemsPKS.pdf).

Compton, P., P. Preston, G. Edwards and B. Kang. 1996. "Knowledge based systems that have some idea of their limits." *Proceedings of the 10th AAAI-Sponsored Banff Knowledge Acquisition for Knowledge-Based Systems Workshop*, Banff. edited by B. Gaines and M. Musen, 50.1–50.18: SRDG Publications, University of Calgary.

Compton, P., P. Preston and B. Kang. 1995. "The Use of Simulated Experts in Evaluating Knowledge Acquisition." *Proceedings of the 9th AAAI-Sponsored Banff Knowledge Acquisition for Knowledge-Based Systems Workshop*, Banff. edited by B. Gaines and M. Musen, 12.1–12.18: SRDG Publications, University of Calgary.

Compton, P., Z. Ramadan, P. Preston, T. Le-Gia, V. Chellen and M. Mullholland. 1998. "A trade-off between domain knowledge and problem-solving method power." *11th Banff Knowledge Acqusition for Knowledge-Based Systems Workshop*, Banff. edited by B. Gaines and M. Musen, SHARE 17, 1–19: SRDG Publications, University of Calgary.

Compton, P. and D. Richards. 2000. "Generalising Ripple-Down Rules." *Knowledge Engineering and Knowledge Management (12th International*

Conference, EKAW 2000). edited by R. Dieng and O. Corby, 380–386: Springer, Berlin.

Crawford, E., J. Kay and E. McCreath. 2002. "IEMS – the intelligent mail sorter." *Machine Learning, Proceedings of the Nineteenth International Conference (ICML 2002)*, Sydney. edited by C. Sammut and A. Hoffmann, 83–90: Morgan Kaufmann.

Cunningham, H., D. Maynard, K. Bontcheva and V. Tablan. 2002. "GATE: A framework and graphical development environment for robust NLP tools and applications." *Proceedings of the 40th Annual Meeting of the Association for Computational Linguistics* Philadelphia, 168–175: The Association for Computational Linguistics.

Curtain, C., I. Bindoff, J. Westbury and G. Peterson. 2013. "An investigation into drug-related problems identifiable by commercial medication review software." *The Australasian Medical Journal* 6 (4):183–188.

D'Este, C. and C. Sammut. 2008. "Learning and generalising semantic knowledge from object scenes." *Journal of Robotics and Autonomous Systems* 56 (11):891–900.

Dani, M. N., T. A. Faruquie, R. Garg, G. Kothari, M. K. Mohania, K. H. Prasad, L. V. Subramaniam and V. N. Swamy. 2010. "Knowledge Acquisition Method for Improving Data Quality in Services Engagements." *IEEE International Conference on Services Computing (SCC 2010)*, Miami, 346–353: IEEE.

Dazeley, R. and B. H. Kang. 2008. "Detecting the knowledge boundary with prudence analysis." *AI 2008: Advances in Artificial Intelligence, 21st Australasian Joint Conference on Artificial Intelligence* edited by W. Wobcke and M. Zhang, 482–488: Springer.

Dazeley, R., S. S. Park and B. H. Kang. 2011. "Online knowledge validation with prudence analysis in a document management application." *Expert Systems With Applications* 38 (9):10959–10965.

Doyle-Lindrud, S. 2015. "Watson will see you now: a supercomputer to help clinicians make informed treatment decisions." *Clinical Journal of Oncology Nursing* 19 (1):31–32.

Drake, B. and G. Beydoun. 2000. "Predicate logic-based incremental knowledge acquisition." *Proceedings of the 6th Pacific International Knowledge Acquisition Workshop*, Sydney. edited by P. Compton, A. Hoffmann, H. Motoda and T. Yamaguchi, 71–88.

Dubinsky, B. and C. Warner. 2008. "Uncovering Accounts Payable Fraud Using Fuzzy Matching Logic: Part 1." *Business Credit* 110 (3):6–9.

Edwards, G., P. Compton, R. Malor, A. Srinivasan and L. Lazarus. 1993. "PEIRS: a pathologist maintained expert system for the interpretation of chemical pathology reports." *Pathology* 25:27–34.

Edwards, G., B. Kang, P. Preston and P. Compton. 1995. "Prudent expert systems with credentials: Managing the expertise of decision support systems." *International Journal of Biomedical Computing* 40:125–132.

Ellis, G. R. 2005. "Improved Search Engine." Ivis, Patent No WO/2005/010775

Eriksson, H., Y. Shahar, S. W. Tu, A. R. Puerta and M. A. Musen. 1995. "Task modeling with reusable problem-solving methods." *Artificial Intelligence* 79 (2):293–326.

Ewings, E. L., P. Konofaos and R. D. Wallace. 2017. "Variations in current procedural terminology coding for craniofacial surgery: a need for review and change." *Journal of Craniofacial Surgery* 28 (5):1224–1228.

Feigenbaum, E. A. 1984. "Knowledge engineering." *Annals of the New York Academy of Sciences* 426 (1):91–107.

Finlayson, A. and P. Compton. 2010. "Incremental knowledge acquisition using generalised RDR for soccer simulation." *Knowledge Management and Acquisition for Smart Systems and Services. The Pacific Knowledge Acquisition Workshop PKAW 2010*, Daegu, Korea. edited by B. Kang and D. Richards, 135–149: Springer.

Finlayson, A. and P. Compton. 2013. "Run-time validation of knowledge-based systems." *Proceedings of the Seventh International Conference on Knowledge Capture (KCAP 2013)*, 25–32: ACM.

Finlayson, A. and P. Compton. 2014. "Using a Domain Expert in Semi-supervised Learning." *Knowledge Management and Acquisition for Smart Systems and Services. Proceedings of the 13th Pacific Rim Knowledge Acquisition Workshop*, Gold Coast, Australia. edited by Y. S. Kim, B. H. Kang and D. Richards, 99–111: Springer.

Forgy, C. 1982. "Rete: A fast algorithm for the many pattern/many object pattern match problem." *Artificial intelligence* 19:17–37.

Forgy, C. and J. McDermott. 1977. "OPS, a domain-independent production system language." *Proceedings of the International Joint Conference on Artificial Intelligence (IJCAI)* 77 (1):933–939.

Forsythe, D. E. 1993. "Engineering knowledge: The construction of knowledge in artificial intelligence." *Social Studies of Science* 23 (3):445–477.

Fox, M. 1990. "AI and expert system myths, legends, and facts." *IEEE EXPERT* 5 (1):8–20.

Frank, E., M. Hall, G. Holmes, R. Kirkby, B. Pfahringer, I. H. Witten and L. Trigg. 2005. *Weka Data Mining and Knowledge Discovery Handbook*. Edited by O. Maimon and L. Rokach: Springer

Gabbay, D. M. 1996. *Labelled deductive systems Vol 1, Oxford Logic Guides*. Oxford: Clarendon Press.

Gaines, B. 1989. "An ounce of knowledge is worth a ton of data: quantitative studies of the trade-off between expertise and data based on statistically well-founded empirical induction." *Proceedings of the Sixth International Workshop on Machine Learning*, San Mateo, California, 156–159: Morgan Kaufmann.

Gaines, B. 1991a. "Induction and visualisation of rules with exceptions." *6th AAAI Knowledge Acquisition for Knowledge Based Systems Workshop*, Banff. edited by J. Boose and B. Gaines, 7.1–7.17: SRDG Publications, University of Calgary.

Gaines, B. 2013. "Knowledge Acquisition: Past, Present and Future." *International Journal of Human Computer Studies* 71 (2):135–156.

Gaines, B. and P. Compton. 1995. "Induction of Ripple-Down Rules Applied to Modeling Large Databases." *Journal of Intelligent Information Systems* 5 (3):211–228.

Gaines, B. and M. Shaw. 1993. "Knowledge acquisition tools based on personal construct psychology." *The Knowledge Engineering Review* 8 (1):49–85.

Gaines, B. R. 1991b. "Integrating Rules in Term Subsumption Knowledge Representation Servers." *AAAI'91: Proceedings of the Ninth National Conference on Artificial Intelligence*, 458–463: AAAI.

Gaines, B. R. and P. J. Compton. 1992. "Induction of Ripple Down Rules." *AI '92. Proceedings of the 5th Australian Joint Conference on Artificial Intelligence*, Hobart, Tasmania. edited by A. Adams and L. Sterling, 349–354: World Scientific, Singapore.

Ganter, B. and R. Wille. 2012. *Formal concept analysis: mathematical foundatiosns*. Berlin: Springer.

Grosof, B. 1997. "Prioritized conflict handling for logic programs." *Logic Programming: Proceedings of the 1997 International Symposium*, 197–211.

Grosof, B. 2004. "Representing e-commerce rules via situated courteous logic programs in RuleMl." *Electronic Commerce Research and Applications* 3:2–20.

Haixiang, G., L. Yijing, J. Shang, G. Mingyun, H. Yuanyue and G. Bing. 2017. "Learning from class-imbalanced data: Review of methods and applications." *Expert Systems with Applications* 73:220–239.

Hall, M., E. Frank, G. Holmes, B. Pfahringer, P. Reutemann and I. H. Witten. 2009. "The WEKA data mining software: an update." *ACM SIGKDD Explorations Newsletter* 11 (1):10–18.

Herbert, D. 2020 - to be submitted. "Conversational knowledge acquisition for conversational agents by non-experts,." PhD, University of Tasmania.

Herbert, D. and B. H. Kang. 2018. "Intelligent conversation system using multiple classification ripple down rules and conversational context." *Expert Systems with Applications* 112:342–352.

Ho, V., W. Wobcke and P. Compton. 2003. "EMMA: an e-mail management assistant." *IAT 2003. IEEE/WIC International Conference on Intelligent Agent Technology*, Halifax. edited by J. Liu, B. Faltings, N. Zhong, R. Lu and T. Nishida, 67–74: IEEE.

Ho, V. H., P. Compton, B. Benatallah, J. Vayssière, L. Menzel and H. Vogler. 2009. "An Incremental Knowledge Acquisition Method for Improving Duplicate Invoice Detection." *Proceedings of the 25th IEEE International Conference on Data Engineering, ICDE 2009*, Shanghai. edited by Y. E. Ioannidis, D. L. Lee and R. T. Ng, 1415–1418 IEEE.

Hoffmann, A. G. 1990. "General Limitations on Machine Learning." *ECAI 90: Proceedings of the 9th European Conference on Artificial Intelligence*, Stockholm. edited by L. C. Aiello, 345–347: Pitman.

Holmes, G., A. Donkin and I. H. Witten. 1994. "WEKA: a machine learning workbench." *Intelligent Information Systems*:357–361.

Hong, C., J. Yu, J. Wan, D. Tao and M. Wang. 2015. "Multimodal deep autoencoder for human pose recovery." *IEEE Transactions on Image Processing* 24 (12):5659–5670.

Horn, K., P. J. Compton, L. Lazarus and J. R. Quinlan. 1985. "An expert system for the interpretation of thyroid assays in a clinical laboratory." *Australian Computer Journal* 17 (1):7–11.

Huang, J., Y. Cai, J. Li, X. Chen and J. Fan. 2019. "Toward Intelligent Train Driving through Learning Human Experience." *2019 1st International Conference on Industrial Artificial Intelligence (IAI)*, Shenyang, 1–6: IEEE.

IBM Research. 2012. *Global Technology Outlook*: IBM

Jha, S. K., Z. Pan, E. Elahi and N. Patel. 2019. "A comprehensive search for expert classification methods in disease diagnosis and prediction." *Expert Systems* 36 (1):e12343.

Kahneman, D., P. Slovic and A. Tversky. 1982. *Judgment under uncertainty: Heuristics and biases*: Cambridge: Cambridge University Press.

Kang, B., W. Gambetta and P. Compton. 1996. "Validation and Verification with Ripple Down Rules." *International Journal of Human Computer Studies* 44 (2):257–270.

Kang, B. H. 2010. "Information Monitoring System on World Wide Web." *Defense Technical Information Center*, Fort Belvoir, Virginia

Kang, B. H., P. Compton, H. Motoda, Y. S. Kim and S. S. Park. 2006. "Towards Next Generation WWW: Push, Reuse and Classification." Storming Media. AOARD-054033

Kang, B. H. and Y. S. Kim. 2009. "Monitoring Web Resources Discovery by Reusing Classification Knowledge." In *Social Computing and Behavioral Modeling*, edited by H. Liu, J. Salerno and M. Young, 1–8. Springer.

Kang, B. H., K. Yoshida, H. Motoda and P. Compton. 1997. "Help desk system with intelligent interface." *Journal of Applied Artificial Intelligence* 11 (7–8):611–631.

Kerr, J. and P. Compton. 2002. "Interactive Learning when Human and Machine Utilise Different Feature Spaces." *The 2002 Pacific Rim Knowledge Acquisition Workshop*, Tokyo. edited by T. Yamaguchi, A. Hoffmann, H. Motoda and P. Compton, 15–29: JSAI.

Kerr, J. and P. Compton. 2003. "Toward Generic Model-based Object Recognition by Knowledge Acquisition and Machine Learning." *Proceedings of the IJCAI-2003 Workshop on Mixed-Initiative Intelligent Systems*, Acapulco. edited by G. Tecuci, D. Aha, M. Boicu, M. Cox, G. Ferguson and A. Tate, 80–86: http://lalab.gmu.edu/miis/proceedings.html.

Kim, D., S. C. Han, Y. Lin, B. H. Kang and S. Lee. 2018. "RDR-based knowledge based system to the failure detection in industrial cyber physical systems." *Knowledge-Based Systems* 150:1–13.

Kim, M. and P. Compton. 2012. "Improving Open Information Extraction for Informal Web Documents with Ripple-Down Rules." *Knowledge Management and Acquisition for Intelligent Systems (PKAW 2012)*, Kuching. edited by D. Richards and B. H. Kang, 160–174: Springer.

Kim, M., P. Compton and Y. S. Kim. 2011. "RDR-Based Open IE for the Web Document." *Proceedings of the Sixth International Conference on Knowledge Capture, KCAP 2011*, Banff. edited by M. Musen and Ó. Corcho, 105–112: ACM.

Kim, M. M. H. 2017. "Incremental Knowledge Acquisition Approach for Information Extraction on both Semi-Structured and Unstructured Text from the Open Domain Web." *Proceedings of the Australasian Language Technology Association Workshop 2017*, Brisbane. edited by J. S.-M. Wong and G. Haffari, 88–96: Australasian Language Technology Association.

Kim, Y., P. Compton and B. Kang. 2012. "Ripple-Down Rules with Censored Production Rules "*Knowledge Management and Acquisition for Intelligent Systems - 12th Pacific Rim Knowledge Acquisition Workshop, PKAW 2012*, Kuching. edited by D. Richards and B. H. Kang, 175–187: Springer.

Kim, Y. S., S. W. Kang, B. H. Kang and P. Compton. 2009. "Using Knowledge Base for Event-Driven Scheduling of Web Monitoring Systems." *E-Commerce and Web Technologies, Proceedings of the 10th International Conference, EC-Web 2009*, Linz. edited by T. Di Noia and F. Buccafurri, 169–180: Springer.

Kim, Y. S., S. S. Park, E. Deards and B. H. Kang. 2004a. "Adaptive web document classification with MCRDR." *International Conference on Information Technology: Coding and Computing (ITCC 2004)*, Las Vegas, (1) 476–480: IEEE.

Kim, Y. S., S. S. Park, B. H. Kang and Y. J. Choi. 2004b. "Incremental Knowledge Management of Web Community Groups on Web Portals." *International Conference on Practical Aspects of Knowledge Management*, Vienna. edited by D. Karagiannis and U. Reimer, 198–207: Springer.

Kivinen, J., H. Mannila and E. Ukkonen. 1994. "Learning rules with local exceptions." *Proceedings Euro-COLT 93*, London. edited by J. Shawe-Taylor and M. Anthony, 35–46: Clarendon Press.

Kolodner, J. 2014. *Case-based reasoning*: San Mateo: Morgan Kaufmann.

Korb, P. J., S. J. Scott, A. C. Franks, A. Virapongse and J. R. Simpson. 2016. "Coding and billing issues in hospital neurology compensation." *Neurology, Clinical Practice* 6 (6):487–497.

Krzywicki, A. and W. Wobcke. 2009. "Incremental e-mail classification and rule suggestion using simple term statistics." *AI 2009: Advances in Artificial Intelligence, 22nd Australasian Joint Conference on Artificial Intelligence*, Melbourne. edited by A. Nicholson and X. Li, 250–259: Springer.

Kuhn, T. 1962. *The structure of scientific revolutions*. Chicago: The University of Chicago Press.

Langley, P. and H. A. Simon. 1995. "Applications of machine learning and rule induction." *Communications of the ACM* 38 (11):54–64.

Lenat, D. B., M. Prakash and M. Shepherd. 1985. "CYC: Using common sense knowledge to overcome brittleness and knowledge acquisition bottlenecks." *AI Magazine* 6 (4):65–85.

Li, G., J. Wang, Y. Zheng and M. J. Franklin. 2016. "Crowdsourced data management: A survey." *IEEE Transactions on Knowledge and Data Engineering* 28 (9):2296–2319.

Lonergan, B. 1959. *Insight*. London: Darton, Longman and Todd.

Loucks, J., T. Davenport and D. Schatsky. 2018. *State of AI in the Enterprise*, 2nd Edition. Deloitte Insights.

Maruatona, O. 2013. "Internet Banking Fraud Detection Using Prudent Analysis." PhD Thesis, University of Ballarat.

Maruatona, O., P. Vamplew and R. Dazeley. 2012a. "Prudent fraud detection in Internet banking." *Third Cybercrime and Trustworthy Computing Workshop (CTC)*, Ballarat, 60–65: IEEE.

Maruatona, O., P. Vamplew and R. Dazeley. 2012b. "RM and RDM, a Preliminary Evaluation of Two Prudent RDR Techniques." *Pacific Knowledge Acquisition Workshop: Knowledge Management and Acquisition for Intelligent Systems*, 188–194: Springer.

Maruatona, O., P. Vamplew, R. Dazeley and P. A. Watters. 2017. "Evaluating accuracy in prudence analysis for cyber security." *Neural Information Processing, 24th International Conference, ICONIP 2017*, Guangzhou. edited by D. Liu, S. Xie, Y. Li, D. Zhao and E. M. El-Alfy, 407–417: Springer.

McCreath, E., J. Kay and E. Crawford. 2006. "IEMS - an approach that combines handcrafted rules with learnt instance based rules." *Australian Journal of Intelligent Information Processing Systems* 9 (1):40–53.

McDermott, J. 1982. "R1: a rule based configurer of computer systems." *Artificial Intelligence* 19 39–88.

Michalski, R. S. and P. H. Winston. 1986. "Variable precision logic." *Artificial intelligence* 29 (2):121–146.

Misra, A., A. Sowmya and P. Compton. 2006. "Incremental Learning for Segmentation in Medical Images." *Proceedings of the 2006 IEEE International Symposium on Biomedical Imaging: From Nano to Macro*, Arlington, 1360–1363: IEEE.

Misra, A., A. Sowmya and P. Compton. 2010. "Incremental system engineering using process networks." *Knowledge Management and Acquisition for Smart Systems and Services. The Pacific Knowledge Acquisition Workshop PKAW 2010*, Daegu, Korea. edited by B. Kang and D. Richards, 150–164: Springer-Verlag.

Motta, E. 1999. *Reusable Components for Knowledge Modelling: Case Studies in Parametric Design Problem Solving*. Amsterdam: IOS press.

Muggleton, S., ed. 1992. *Inductive Logic Programming*. London: Academic Press.

Mulholland, M., D. Hibbert, P. Haddad and P. Parslov. 1995. "A comparison of classification in artificial intelligence, induction versus a self-organising neural networks." *Chemometrics and intelligent laboratory systems* 30 (1):117–128.

Newell, A. and H. A. Simon. 1976. "Computer science as empirical inquiry: symbols and search." *Communications of the ACM* 19 (3):113–126.

Nguyen, D. B. and S. B. Pham. 2012. "Ripple down rules for Vietnamese named entity recognition." *Computational Collective Intelligence. Technologies and Applications, 4th International Conference, ICCCI 2012*, Ho Chi Minh City. edited by N.-T. Nguyen, K. Hoang and P. Jędrzejowicz, 354–363: Springer.

Nguyen, D. Q., D. Q. Nguyen, D. D. Pham and S. B. Pham. 2016. "A robust transformation-based learning approach using ripple down rules for part-of-speech tagging." *AI Communications* 29 (3):409–422.

Nguyen, D. Q., D. Q. Nguyen and S. B. Pham. 2017a. "Ripple down rules for question answering." *Semantic Web* 8 (4):511–532.

Nguyen, D. Q., D. Q. Nguyen, S. B. Pham and D. D. Pham. 2011. "Ripple down rules for part-of-speech tagging." *Proceedings of the 12th international conference on Computational linguistics and intelligent text processing* Tokyo, 190–201: Springer.

Nguyen, D. Q., D. Q. Nguyen, T. Vu, M. Dras and M. Johnson. 2017b. "A Fast and Accurate Vietnamese Word Segmenter." *arXiv preprint* 1709.06307.

OMG. 2009. "Production Rule Representation (PRR)." *Object Management Group* http://www.omg.org/spec/PRR/1.0.

Park, S. S., Y. S. Kim and B. H. Kang. 2003. "Web Information Management System: Personalization and Generalization." *IADIS International Conference WWW/Internet*, Algarve. edited by P. Isaías and N. Karmakar, 523–530: IADIS.

Park, S. S., Y. S. Kim and B. H. Kang. 2004. "Web Document Classification: Managing Context Change." *IADIS International Conference WWW/Internet*, Madrid. edited by P. Isaías and N. Karmakar, 143–151: IADIS.

Pérez, A. G. and V. R. Benjamins. 1999. "Overview of knowledge sharing and reuse components: Ontologies and problem-solving methods." *Proceedings of the IJCAI-99 workshop on Ontologies and Problem-Solving methods (KRR5)*, Stokholm, 1–15: CEUR Publications.

Pham, K. C. and C. Sammut. 2005. "RDRVision - learning vision recognition with ripple down rules." *Proc. Australasian Conference on Robotics and Automation*, https://www.researchgate.net/publication/228967796_Rdrvision-learning_vision_recognition_with_ripple_down_rules.

Pham, S. B. and A. Hoffmann. 2004. "Extracting Positive Attributions from Scientific Papers." *Discovery Science: 7th International Conference, DS 2004*, Padova. edited by E. Suzuki and S. Arikawa, 169–182: Springer.

Pham, S. B. and A. Hoffmann. 2006. "Efficient Knowledge Acquisition for Extracting Temporal Relations." *ECAI 2006: 17th European Conference on Artifcial Intelligence* Riva del Garda, 354–359: IOS Press.

Polit, S. 1984. "R1 and Beyond: AI Technology Transfer at Digital Equipment Corporation." *AI Magazine* 5 (4):76–78.

Popper, K. 1963. *Conjectures and refutations*. London: Routledge and Kegan Paul.

Prayote, A. 2007. "Knowledge based anomaly detection." Ph.D., University of New South Wales, Sydney, Australia.

Prayote, A. and P. Compton. 2006. "Detecting Anomalies and Intruders." *AI 2006: Advances In Artificial Intelligence, 19th Australia Joint Conference on Artifical Intelligence*, Hobart, Australia. edited by A. Sattar and B. Kang, 1084–1088: Springer.

Preston, P., G. Edwards and P. Compton. 1994. "A 2000 Rule Expert System Without a Knowledge Engineer." *Proceedings of the 8th AAAI-Sponsored Banff Knowledge Acquisition for Knowledge-Based Systems Workshop*, Banff. edited by B. Gaines and M. Musen, 17.1–17.10: SRDG Publications, University of Calgary.

Puppe, F. 1993. *Systematic introduction to expert systems: Knowledge representations and problem-solving methods*. Berlin: Springer-Verlag.

Quinlan, J. 1992. *C4.5: Programs for Machine Learning*: San Francisco: Morgan Kaufmann.

Ramadan, Z., M. Mulholland, D. Hibbert, P. Preston, P. Compton and P. Haddad. 1998a. "Towards an expert system in ion exclusion chromatography using multiple classification ripple down rules (RDR)." *Journal of Chromatography* 795:29–35.

Ramadan, Z., P. Preston, P. Compton, T. Le-Gia, V. Chellen, M. Mullholland, D. Hibbert, P. Haddad and B. Kang. 1998b. "From Multiple Classification RDR to Configuration RDR." *11th Banff Knowledge Acqusition for Knowledge-Based Systems Workshop*, Banff. edited by B. Gaines and M. Musen, SHARE 17, 1–19: SRDG Publications, University of Calgary.

Rashidi, H. H., N. K. Tran, E. V. Betts, L. P. Howell and R. Green. 2019. "Artificial intelligence and machine learning in pathology: the present landscape of supervised methods." *Academic Pathology* 6: 2374289519873088.

Richards, D. 2009. "Two decades of Ripple Down Rules research." *The Knowledge Engineering Review* 24 (2):159–184.

Richards, D. and P. Compton. 1997. "Combining Formal Concept Analysis and Ripple Down Rules to Support Reuse." *Software Engineering and Knowledge Engineering SEKE'97*, Madrid, 177–184: Springer.

Richards, D. and P. Compton. 1999. "Revisiting Sisyphus I - an Incremental Approach to Resource Allocation Using Ripple-Down Rules." *12th Banff Knowledge Acqusition for Knowledge-Based Systems Workshop*, Banff. edited by B. Gaines, R. Kremer and M. Musen, 7-7.1–7-7.20: SRDG Publications, University of Calgary.

Rosenbloom, S. T., R. A. Miller, K. B. Johnson, P. L. Elkin and S. H. Brown. 2006. "Interface Terminologies: Facilitating Direct Entry of Clinical Data into Electronic Health Records." *Journal of the American Medical Informatics Association* 13:277–288.

Ryu, S. H. and B. Benatallah. 2012. "Integrating feature analysis and background knowledge to recommend similarity functions." *Web Information Systems Engineering - WISE 2012*, Paphos. edited by X. S. Wang, I. Cruz, A. Delis and G. Huang, 673–680: Springer.

Ryu, S. H. and B. Benatallah. 2017. "Experts community memory for entity similarity functions recommendation." *Information Sciences* 379:338–355.

Ryu, S. H., B. Benatallah, H.-Y. Paik, Y. S. Kim and P. Compton. 2011. "Similarity function recommender service using incremental user knowledge acquisition." *Service-Oriented Computing 9th International Conference, ICSOC 2011*, Paphos. edited by G. Kappel, Z. Maamar and H. Motahari-Nezhad, 219–234: Springer.

Sarraf, Q. and G. Ellis. 2006. "Business Rules in Retail: The Tesco.com Story." *Business Rules Journal* 7 (6):http://www.BRCommunity.com/a2006/n014.html.

Schank, R. C. and R. P. Abelson. 1977. *Scripts, Plans, Goals and Understanding: An Enquiry into Human Knowledge Structures*. Hillsdale, New Jersey: Erlbaum.

Scheffer, T. 1996. “Algebraic foundations and improved methods of induction of ripple-down rules.” *PKAW’96: The Pacific Knowledge Acquisition Workshop*, Sydney. edited by P. Compton, R. Mizoguchi, H. Motoda and T. Menzies, 279–292.

Schreiber, G., H. Akkermans, A. Anjewierden, R. de Hoog, N. Shadbolt, W. van de Velde and B. Wielinga. 1999. *Knowledge Engineering and Management: The CommonKADS Methodology*. Cambridge, Mass.: MIT Press.

Schreiber, G., B. Wielinga, R. D. Hoog, H. Akkermans and W. van de Velde. 1994. “CommonKADS: A comprehensive methodology for KBS development.” *IEEE Expert* 9 (6):28–37.

Shadbolt, N. and P. Smart. 2015. “Knowledge Elicitation.” In *Evaluation of Human Work* (4th ed.), edited by J. R. Wilson and S. Sharples, 163–200. Boca Raton: CRC Press.

Shaw, M. and J. Woodward. 1988. “Validation in a knowledge support system: construing and consistency with multiple experts.” *International Journal of Man-Machine Studies* 29:329–350.

Shaw, M. L. G. 1980. *On Becoming a Personal Scientist: Interactive Computer Elicitation of Personal Models of the World*: London: Academic Press, Inc.

Shiraz, G. and C. Sammut. 1997. “Combining Knowledge Acquisition and Machine Learning to Control Dynamic Systems.” *15th International Joint Conferences on Artificial Intelligence (IJCAI’97)* Nagoya. edited by M. E. Pollack, 908–913: Morgan Kaufmann.

Shortliffe, E. H., A. C. Scott, M. B. Bischoff, W. van Melle and C. D. Jacobs. 1981. “ONCOCIN: an expert system for oncology protocol management.” *Proceedings of the 7th International Joint Conference on Artificial Intelligence*, Vancouver, 876–881: Morgan Kauffman.

Siromoney, A. and R. Siromoney. 1993. “Local exceptions in inductive logic programming.” In *Machine intelligence 14: Applied Machine Intelligence*, edited by K. Furukawa, D. Michie and S. Muggleton, 211–232. Oxford University Press.

Soloway, E., J. Bachant and K. Jensen. 1987. “Assessing the maintainability of XCON-in-RIME: coping with the problems of a VERY large rule base.” *Proceedings of AAAI 87*, Seattle, 824–829: Morgan-Kauffman.

Speel, P., A. T. Schreiber, W. van Joolingen, G. van Heijst and G. Beijer. 2001. “Conceptual Models for Knowledge-Based Systems.” In *Encyclopedia of Computer Science and Technology*, edited by A. Kent and J. G. Williams, 107–132. Marcel Dekker.

Spiegelhalter, D. 2020. “Should we trust algorithms?” *Harvard Data Science Review* 2 (1). doi:10.1162/99608f92.cb91a35a

Sridhar, A. 2011. “Combining Computer Vision and Knowledge Acquisition to provide Real-Time Activity Recognition for Multiple Persons within Immersive Environments.” PhD, The University of New South Wales.

Sridhar, A., A. Sowmya and P. Compton. 2010. “On-line, incremental learning for real-time vision based movement recognition.” *The Ninth International Conference on Machine Learning and Applications (ICMLA 2010)*, Washington, DC, 465–470: IEEE.

Strickland, E. 2019. “IBM Watson, heal thyself: How IBM overpromised and underdelivered on AI health care.” *IEEE Spectrum* 56 (04):24–31.

Suryanto, H. and P. Compton. 2002. "Intermediate Concept Discovery in Ripple Down Rule Knowledge Bases." *The 2002 Pacific Rim Knowledge Acquisition Workshop (PKAW 2002)*, Tokyo. edited by T. Yamaguchi, A. Hoffmann, H. Motoda and P. Compton, 233–245.

Suryanto, H. and P. Compton. 2003. "Invented Predicates to Reduce Knowledge Acquisition Effort." *Proceedings of the IJCAI-2003 Workshop on Mixed-Initiative Intelligent Systems*, Acapulco. edited by G. Tecuci, D. Aha, M. Boicu, M. Cox, G. Ferguson and A. Tate, 107–114: http://lalab.gmu.edu/miis/proceedings.html.

Suryanto, H. and P. Compton. 2004. "Invented Predicates to Reduce Knowledge Acquisition." *Engineering Knowledge in the Age of the Semantic Web (EKAW 2004)*: Whittleburg Hall, UK. edited by E. Motta and N. Shadbolt, 293–306: Springer.

Suryanto, H., D. Richards and P. Compton. 1999. "The Automatic Compression of Multiple Classification Ripple Down Rule Knowledge Base Systems: Preliminary Experiments." *Proceedings of the Third International Conference on Knowledge-Based Intelligent Information Engineering Systems*, Adelaide. edited by L. Jain, 203–206: IEEE.

Sviokla, J. 1990. "An examination of the impact of expert systems on the firm: the case of XCON." *Management Information Systems Quarterly* 14 (2):127–140.

ten Teije, A., F. van Harmelen, A. T. Schreiber and B. J. Wielinga. 1998. "Construction of problem-solving methods as parametric design." *International Journal of Human-Computer Studies* 49 (4):363–389.

Tordai, A., J. van Ossenbruggen, G. Schreiber and B. Wielinga. 2011. "Let's agree to disagree: on the evaluation of vocabulary alignment." *Proceedings of the Sixth International Conference on Knowledge Capture, KCAP 2011*, Banff, 65–72: ACM.

Vazey, M. and D. Richards. 2004. "Achieving rapid knowledge acquisition in a high-volume call centre." *Proceedings of the Pacific Knowledge Acquisition Workshop 2004*, Auckland. edited by B. H. Kang, A. Hoffmann, T. Yamaguchi and W. K. Yeap, 74–86: University of Tasmania Eprints Repository.

Vazey, M. and D. Richards. 2005. "Troubleshooting at the Call Centre: A Knowledge-based Approach." *Artificial Intelligence and Applications*, 721–726.

Vazey, M. and D. Richards. 2006a. "A case-classification-conclusion 3Cs approach to knowledge acquisition: Applying a classification logic Wiki to the problem solving process." *International Journal of Knowledge Management* 2 (1):72–88.

Vazey, M. and D. Richards. 2006b. "Evaluation of the FastFIX Prototype 5Cs CARD System." *Advances in Knowledge Acquisition and Management, PKAW 2006*, Guilin. edited by A. Hoffmann, B.-H. Kang, D. Richards and S. Tsumoto, 108–119: Springer.

Vazey, M. M. 2007. "Case-driven collaborative classification." PhD, Department of Computing, Division of Information and Communication Sciences, Macquarie University.

Wada, T., T. Horiuchi, H. Motoda and T. Washio. 2000. "Integrating inductive learning and knowledge acquisition in the Ripple Down Rules method."

Proceedings of the 6th Pacific International Knowledge Acquisition Workshop, Sydney. edited by P. Compton, A. Hoffmann, H. Motoda and T. Yamaguchi, 268–278.

Wada, T., T. Horiuchi, H. Motoda and T. Washio. 2001. "A description length-based decision criterion for default knowledge in the Ripple Down Rules method." *Knowledge and Information Systems* 3 (2):146–167.

Wagner, W. P. 2017. "Trends in expert system development: A longitudinal content analysis of over thirty years of expert system case studies." *Expert Systems with Applications*:85–96.

Wai, K. H., Y. K. Thu, H. A. Thant, S. Z. Moe and T. Supnithi. "Myanmar (Burmese) String Similarity Measures based on Phoneme Similarity." *Journal of Intelligent Informatics and Smart Technology* 4 (April 2020):27–34.

Wang, J. C., M. Boland, W. Graco and H. He. 1996. "Use of ripple-down rules for classifying medical general practitioner practice profiles repetition." *Proceedings of Pacific Knowledge Acquisition Workshop PKAW'96*. edited by P. Compton, R. Mizoguchi, H. Motoda and T. Menzies, 333–345.

Waterman, D. 1986. *A guide to expert systems*. Reading, MA: Addison Wesley.

Winograd, T. and F. Flores. 1987. *Understanding computers and cognition*. Reading, MA: Addison Wesley.

Wobcke, W., A. Krzywicki and Y.-W. Chan. 2008. "A Large-Scale Evaluation of an E-Mail Management Assistant." *Proceedings of the 2008 IEEE/WIC/ACM International Conference on Web Intelligence and Intelligent Agent Technology*, 438–442: IEEE

Xu, H. and A. Hoffmann. 2010. "RDRCE: combining machine learning and knowledge acquisition." *Knowledge Management and Acquisition for Smart Systems and Services, PKAW 2010*, Daegu. edited by B. Kang and D. Richards, 165–179: Springer.

Yoshida, T., T. Wada, H. Motoda and T. Washio. 2004. "Adaptive Ripple Down Rules method based on minimum description length principle." *Intelligent Data Analysis* 8 (3): 239–265.

Zacharias, V. 2008. "Development and Verification of Rule Based Systems—A Survey of Developers." *RuleML 2008*, Orlando. edited by N. Bassiliades, G. Governatori and A. Paschke, 6–16: Springer-Verlag.

Zacharias, V. 2009. "The Debugging of Rule Bases." In *Handbook of Research on Emerging Rule-Based Languages and Technologies: Open Solutions and Approaches*, 302–325. IGI Global.

Zaw, W. T., S. S. Moe, Y. K. Thu and N. N. Oo. 2020. "Applying Weighted Finite State Transducers and Ripple Down Rules for Myanmar Name Romanization." *2020 17th International Conference on Electrical Engineering/Electronics, Computer, Telecommunications and Information Technology (ECTI-CON)*, 143–148: IEEE.

Zhou, D., H.-Y. Paik, S. H. Ryu, J. Shepherd and P. Compton. 2016. "Building a process description repository with knowledge acquisition." *Pacific Rim Knowledge Acquisition Workshop*, Phuket. edited by H. Ohwada and K. Yoshida, 86–101: Springer.

Index

W

X

Z

Taylor & Francis eBooks

www.taylorfrancis.com

A single destination for eBooks from Taylor & Francis with increased functionality and an improved user experience to meet the needs of our customers.

90,000+ eBooks of award-winning academic content in Humanities, Social Science, Science, Technology, Engineering, and Medical written by a global network of editors and authors.

TAYLOR & FRANCIS EBOOKS OFFERS:

- A streamlined experience for our library customers
- A single point of discovery for all of our eBook content
- Improved search and discovery of content at both book and chapter level

REQUEST A FREE TRIAL

support@taylorfrancis.com